Ceramic Hydrogen Storage Materials

Ceramic Hydrogen Storage Materials presents the physical, chemical, and physico-chemical properties of ceramics as hydrogen storage materials. It demonstrates how ceramic nanostructures can be specifically designed for high surface adsorption of hydrogen and can function as hydrogen batteries.

Covering methods for characterizing hydrogen storage capacity, this book showcases how hydrogen has the potential to facilitate the decarbonization of the electric power sector by storing energy produced from renewable sources. It addresses both ceramic oxides and non-oxides while expanding on the synthesis and characterization of zero- and one-dimensional, high-surface-area ceramic structures. This book discusses the applications of hydrogen batteries in zero-emission vehicles and hydrogen fuels for aircraft.

This book will interest upper-level undergraduate energy and materials engineering students, as well as researchers studying green energy storage materials, hydrogen storage, and alternative transportation fuels.

Ceramic Hydrogen Storage Materials

High Storage Density Candidates for Sustainable Green Energy

Navid Hosseinabadi

CRC Press
Taylor & Francis Group
Boca Raton London New York

CRC Press is an imprint of the
Taylor & Francis Group, an **informa** business

First edition published 2025
by CRC Press
2385 NW Executive Center Drive, Suite 320, Boca Raton FL 33431

and by CRC Press
4 Park Square, Milton Park, Abingdon, Oxon, OX14 4RN

CRC Press is an imprint of Taylor & Francis Group, LLC

ISBN: 978-1-032-78644-5 (hbk)
ISBN: 978-1-032-78649-0 (pbk)
ISBN: 978-1-003-48882-8 (ebk)

DOI: 10.1201/9781003488828

Typeset in Times
by codeMantra

To my beloved

Naghmeh and Noyan

Contents

Preface

Hydrogen, a green energy source, can play an essential role in the sustainable economy of the coming century. The International Energy Agency has projected that by the year 2050, hydrogen energy will account for 6% of the cumulative emission reductions necessary to achieve global net-zero emissions. While hydrogen is a vital chemical source for numerous industries aimed at decreasing carbon footprints, it must be secured through sustainable sources for all these applications. Hydrogen consumption is primarily proposed through the de-carbonization of the energy sectors, utilizing well-organized and clean power generation via hydrogen fuel cells, and secondly, by investing in hydrogen's high gravimetric density through efficient energy storage mediums for the energy and transportation sectors. The necessary requirements for hydrogen storage materials and hydrogen batteries commonly focus on high storage density and rapid physically and thermally activated adsorption/desorption kinetics. Therefore, the development of hydrogen-sourced industries, especially in the transportation sector, depends on introducing new and advanced hydrogen adsorbents with high durability and operational confidence. To address these concerns, researchers and material development groups in academia and industry have proposed fine oxide and non-oxide ceramics as adsorption mediums.

This volume covers these requirements through Chapter 1 ("Hydrogen Gas: Future Green Energy Source," a brief and concise overview of future perspectives on hydrogen energy sources); Chapter 2 ("Ceramics: Structures, Types, and Properties," a detailed chapter providing general information about different categories of ceramic materials: oxides, non-oxides, monolithics, and composites); Chapter 3 ("Gas Adsorption: Thermodynamics and Kinetics," which presents detailed content on the science of gas adsorption on solid surfaces); Chapter 4 ("Hydrogen Adsorption," which discusses the physisorption of molecular gases with an emphasis on hydrogen); Chapter 5 ("Hydrogen Storage: Ceramic Hydrogen Batteries," a chapter devoted to the development of solid-state surfaces as adsorption sites for hydrogen storage, with examples of oxides, non-oxides, nanoparticles, nanocapsules, MXenes, and similar materials); and Chapter 6 ("Analysis and Assessment: Challenges, Opportunities, and Future"). The ultimate objective of this volume is to provide insights into the latest developments in sustainable hydrogen storage using ceramic materials, addressing both academia and industry, and exploring applications of these technologies.

Navid Hosseinabadi

About the Author

Dr. Navid Hosseinabadi is currently a faculty member at Shiraz University, Iran. He specializes in advanced ceramics science and engineering. His research interests focus on developing solid-state inorganic hydrogen storage materials as hydrogen batteries from ceramics, including nanoparticles, nanocapsules, MXenes, and ceramic substrates.

1 Hydrogen Gas
Future Green Energy Source

Scope

The importance of hydrogen energy as an alternative energy source to fossil-based conventional fuels is gaining ground, especially with the growing concerns for environmental preservation (climate objectives) and the introduction of regulations setting mandatory national targets for the deployment of alternative fuels infrastructure by the European Union, the United States, and the UN. While the necessity of addressing these concerns is clear, the technological aspects need closer attention and targeted research to facilitate hydrogen consumption.

1.1 HYDROGEN: AN EXCEPTIONAL ELEMENT

Hydrogen, the lightest element and a gas of diatomic molecules at standard conditions (dihydrogen), is widely known for its non-toxicity and high combustibility. Most of hydrogen's uniqueness is related to its +1 oxidation state ($H^+ - H^-$), 13.598 eV ionization energy, 2.2 (Pauling scale) electronegativity, 1.08×10^5 milligrams per liter abundance, 0.75–0.77 eV electron affinity, and ~2.01568 g/mole molar mass [1]. Additional basic information on the hydrogen atom/molecule is reported in Table 1.1.

Hydrogen atoms are formed via a nucleus consisting of a proton with one unit of positive electrical charge and an electron with one unit of negative electrical charge. Generally, and under usual conditions, hydrogen gas consists of a loose aggregation of hydrogen molecules; each composed of a pair of atoms in a diatomic molecule. As a classic example of a covalent bond that is based on the exchange

TABLE 1.1
General Information on the Nature of Hydrogen

Atomic radius	0.120 nm
Orbital radius	0.053 nm
Covalent radius	0.037 nm
Positive ion radius	0.84184×10^{-6} nm
Negative ion radius	0.208 nm
Absolute weight	1.67261×10^{-27} kg
Nuclear magnetic moment	+2.7928 μN
Nuclear spin	½ (h/2π)

DOI: 10.1201/9781003488828-1

forces, molecular hydrogen (H_2) is a combination of the two isomers of ortho- and para-hydrogen with different nuclear magnetic moments (spins). While ortho-hydrogen ($O - H_2$) has a similar orientation, that of para-hydrogen ($P - H_2$) has an opposite orientation. Generally, normal hydrogen gas contains nearly 75% of $O - H_2$ and less than 25% of $P - H_2$. The conversion of $O - H_2$ to $P - H_2$ releases heat as much as 1400 kJ/mol. This characteristic signifies the importance of special attention to molecular hydrogen besides atomic hydrogen [2]. Some of the indicative properties of molecular hydrogen are listed in Table 1.2.

The chemical reaction probability of hydrogen atoms and molecules is function of standard thermodynamic properties. Some of these properties are listed in Table 1.3 for H_2, H, H^+, and H^- at ambient temperature (25°C) and pressure (1 atm).

Besides atomic and molecular hydrogen, hydrogen contains three isotopes of protium (^1_1H), deuterium (^2_1H), and tritium (^3_1H). Protium has one electron and one proton, deuterium contains one electron, one proton and one neutron, while tritium has one electron, one proton and two neutrons. Among these, deuterium and tritium isotopes are used in fusion power for electrical generation and require separation of deuterium and tritium in high purity. As isotopes have similar chemical properties, isotope separation is more difficult than chemical separation. A schematic view of

TABLE 1.2

Structural Properties of Hydrogen

Mass (molar)	2.016
Mass (absolute)	3.50×10^{-27} kg
Bond energy	-4.749 eV
Dissociation energy	4.749 eV
Bond length	0.0742 nm
Energy (vibration)	0.5162 eV
Energy (rotation)	0.0073 eV
Density (gas)	0.09 (kg/m³)
Density (liquid)	70.81 (kg/m³)
Viscosity	9.0×10^{-6} kg/ms
Specific volume	12.1 (m³/kg)

TABLE 1.3

Thermodynamic Properties of Hydrogen and Its Oxidation States

	Heat capacity C_P^0 (J/mol K)	Entropy S^0 (J/mol K)	Enthalpy H^0 (kJ/mol)	Enthalpy of formation $\Delta_f H^0$ (kJ/mol)	Energy of formation $\Delta_f G^0$ (kJ/mol)
H	20.79	114.72	6.20	218.00	203.28
H_2	28.84	130.70	8.47	0	0
H^+	20.78	108.95	6.20	1536.25	1516.9
H^-	2079	108.96	6.20	139.03	132.28

hydrogen in its most important forms is shown in Figure 1.1. Beyond structural properties, a combination of thermodynamic properties indicates hydrogen's behavior in operational conditions. Table 1.4 shows some of these characteristics including heat of combustion, critical conditions, and latent heats.

Production of hydrogen generally requires a large quantity of energy. The main precursors of hydrogen production processes include fossil fuels, hydrocarbons, steam, air, and oxygen. Processes are carried out by heating and combining in a reactor. Industrial hydrogen is recovered from both water phases and hydrocarbons, with H_2, CO, and CO_2 being produced as by-products from aqueous

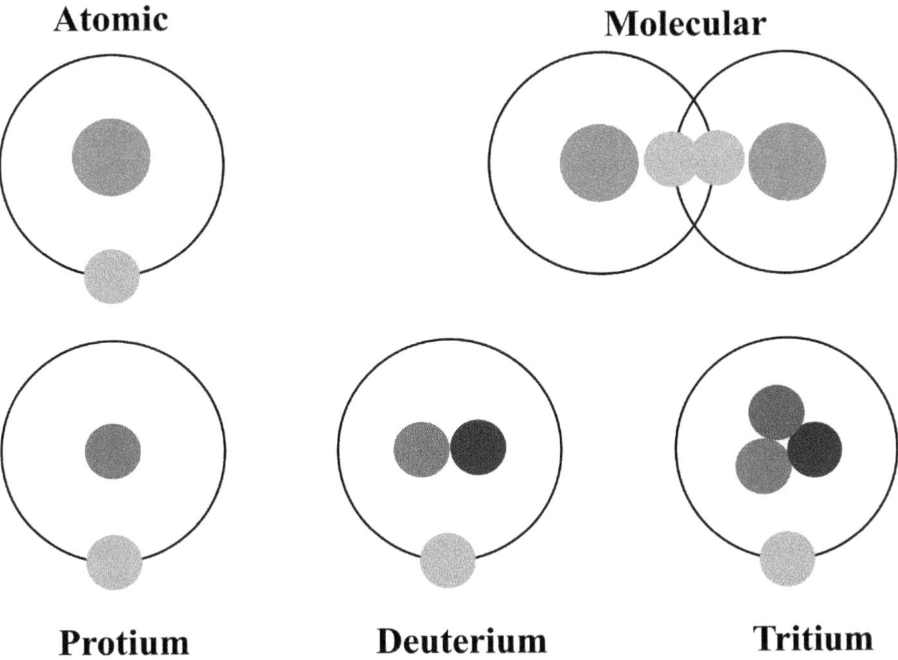

FIGURE 1.1 Schematic view of hydrogen in its most important forms. Created, designed, and introduced by the author.

TABLE 1.4

Thermophysical Properties of Hydrogen

Heat of combustion	144000 kJ/mol
Critical temperature	−240°C
Critical pressure	1.30 MPa
Latent heat of evaporation	447000 kJ/mol
Latent heat of fusion	58 kJ/kg
Specific heat ratio	1.41
Gas constant	4126 J/kg K
Thermal conductivity	0.7–1.1 W/m K

molecules and hydrocarbons. The decomposition of hydrocarbons into hydrogen and carbon in the absence of steam and/or air uses natural gas or refined oil by-products. Recent data shows that half of the hydrogen produced worldwide is derived from natural gas through steam methane reforming, around 30% from oil, and the rest from coal and water electrolysis. Therefore, the main methods are catalytic steam reforming of natural gas, partial oxidation of petroleum, steam iron treatment, and the gasification of coal alongside electrolysis. Hydrogen can be a by-product of many industrial processes like the production of coke from coal in coke ovens, the production of counter-chlorine in the chlor-alkali industry, the production of light gases in the crude oil refining process, and chemical hydrogenation processes in the margarine industry [3]. A short description of some of these production methods is mentioned below.

Production of hydrogen from fossil fuels is an easy and accessible method without complicated infrastructure requirements. The main disadvantage of this process is the CO_2 release of CO as a by-product. Natural gas reforming utilizes natural gas through steam reforming, partial oxidation, and autothermal reforming. The *steam reforming* process is less costly and more efficient. It involves the endothermic conversion of methane, ethanol, propane, and gasoline along with high-temperature steam (at 700°C–1000°C) in the presence of catalysts at pressures of 5–25 bar. The products and by-products of this process are hydrogen, CO, and CO_2. Supplementary processing of carbon monoxide and steam creates additional reactions in the presence of a catalyst, which produces carbon dioxide and extra hydrogen. Through *partial oxidation*, natural gas is exposed to a low concentration of oxygen without completely oxidizing the fuel. Consequently, the gas is partially oxidized, and the products are carbon monoxide and hydrogen. During *autothermal* reforming (water–gas shift reaction), a combination of steam reforming and partial oxidation is used, where carbon monoxide (a by-product of the partial oxidation process) reacts with high-temperature steam to produce extra hydrogen (an exothermic step). As coal is a complex precursor, combinations of many processes including fixed-bed, fluidized-bed, or entrained flow are used for hydrogen production through the *coal gasification* process. During the process, coal initially reacts with oxygen and steam under high pressure. The products of this process are carbon monoxide, CO/CO_2, and hydrogen (following reaction), where the CO by-product undergoes another water–gas shift reaction to produce more hydrogen.

$$C_x H_y O_z NS + O_2 + H_2O \rightarrow CO + CO_2 + H_2 \qquad (1.1)$$

Producing hydrogen from water by *electrolysis* is one of the most energy-intensive ways to produce fuel. Also known as *water splitting*, it is among the most accessible hydrogen production methods by splitting water into its elements. It is carried out through a variety of technologies, including electrolysis or solar thermochemical splitting. While the required energy for electrolysis at higher temperatures significantly increases, the required electrical energy decreases. Accordingly, electrolysis at higher temperatures is a standard practice where waste heat from different resources can be used as a heat source. *High-temperature electrolysis* (high-temperature fuel cell technology) utilizes a high-temperature electrolyzer with higher overall efficiency compared to a low-temperature electrolyzer. In these systems, heat sources like geothermal, solar, or natural gas have replaced electricity to reduce electricity consumption. Although it is a clean process as long as the electricity comes from a clean source, electrolysis is associated with extensive energy losses. Electrolysis is carried out at the standard potential of about 1.23 V. Nevertheless, the electrolyzers require higher voltage to separate water into hydrogen and oxygen. This overpotential is required to overcome polarization and ohmic losses caused by electric current flow under operational conditions. Solar is considered as a source of heat integrated with high-temperature electrolysis to reduce electricity consumption; *photoelectrolysis* (photolysis) is developed to use sunlight directly to split water. *Solar thermochemical hydrogen* is another high-temperature water-splitting technology developed using concentrated solar power, high temperature, or waste heat of a nuclear reactor in combination with chemical reactions to produce hydrogen. Low greenhouse gas emission is the main advantage of this process. Through solar thermochemical hydrogen production, high temperatures are utilized to drive a chemical reaction (using cerium oxide and copper chloride) to split water. Hydrogen production through *photobiological* production (biophotolysis) includes two steps of photosynthesis and hydrogenase-catalyzed production as designed for artificial photosynthesis. A combination of hydrogen production and biogas production is used in the newly developed *biomass conversion* process. During the process, the original resource is converted to a hydrogen-containing gas at high temperatures without combustion, similar to coal gasification. The process incorporates steam gasification, entrained flow gasification, application of thermochemical cycles, and conversion of ethanol and bio-oil by-products [4]. Figure 1.2 illustrates the sources and methods of hydrogen production (volume).

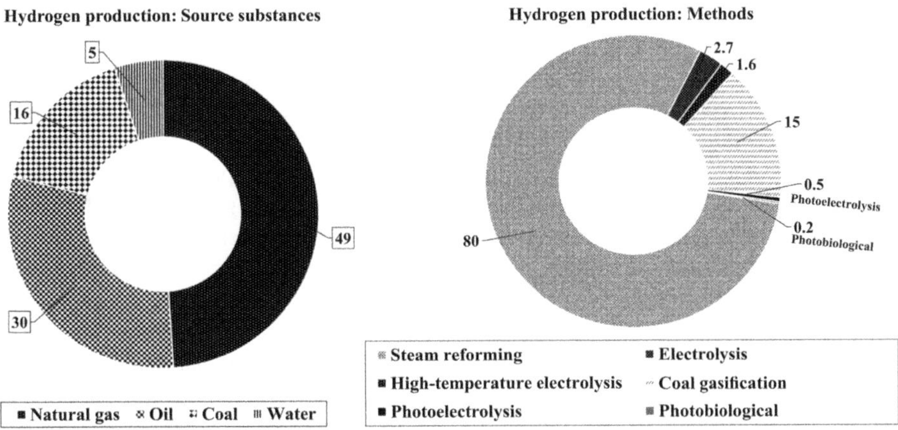

FIGURE 1.2 Hydrogen production sources and methods with estimated share in the current industry. Created, designed, and introduced by the author.

1.2   HYDROGEN: AN ALTERNATIVE ENERGY SOURCE

The swift climate change is a serious issue, and the more or less continuous increase CO_2 in levels has significantly raised concerns about the global warming phenomenon and the rise in global average temperature. In addition, energy loss in the current energy economy amounts to about 10% of the energy delivered to the final consumers. These concerns are the main driving force behind developing alternative energy sources, especially hydrogen.

Hydrogen, in both atomic and molecular forms, is a significant energy carrier that only forms water through combustion with oxygen – a simple and clean heat conversion to power without pollutant generation or emission. Hydrogen is an energy carrier capable of transforming fossil fuel-dependent domestic life and industries into a clean society and economy. It can become the main link between renewable physical energy and chemical energy carriers. Since pure hydrogen does not exist naturally, it must be separated from chemical compounds through methods such as electrolysis from water or chemical processes from hydrocarbons or other hydrogen carriers, such as steam methane reforming, which involves exposing natural gas to heat and steam to separate carbon atoms from methane, forming hydrogen, and gasification of coal/biomass. Additional methods of hydrogen production include photoelectrolysis, where sunlight is used to separate water into hydrogen and oxygen by applying a photochemical cell submerged in water, and thermochemical cycles where water is heated to 900°C to split water molecules and release hydrogen. Moreover, some microorganisms can be used to produce hydrogen through a photosynthetic process using sunlight. Clean electric power for electrolysis can be supplied from renewable sources such as solar radiation, kinetic wind energy, and water/geothermal heat. Hydrogen, as a synthetic energy carrier, contains energy generated through other processes. For example, electrical energy needs to be transferred to hydrogen through water electrolysis. Therefore, hydrogen has emerged as one of the promising candidates for alternative energy sources to reduce the environmental impact of greenhouse gas emissions – regenerative and environmentally friendly. Its easy and approachable production method has made hydrogen a significant alternative energy source to address environmental concerns and meet the increasing energy demands due to the depletion of fossil fuels. Another important aspect of hydrogen as an energy carrier is its exhibition of the highest heating value per mass (39.4 kWh/kg) among all chemical fuels. As a result, hydrogen, being the lightest gas (about eight times lighter than natural gas methane), has an impressive heating value. With a density of 0.09, the gravimetric and volumetric heating values of 142 MJ/kg and 12.7 MJ/m^3 are notable, respectively, indicating the true energy content of the fuel based on the energy conservation principle (first law of thermodynamics). Hydrogen gas is known to have the highest higher heating value per kilogram among other gases [5]. For comparison, 324 g of hydrogen contains the same energy content as 1 kg of gasoline but occupies three times the volume – a disadvantage that can be resolved. At any operating pressure, hydrogen gas contains less energy per volume compared to propane, methane (natural gas), methanol, or octane (gasoline). At high pressures of 800 bar, gaseous hydrogen reaches the volumetric energy density of liquid hydrogen. Otherwise, at any pressure, the volumetric energy density of methane gas exceeds that of hydrogen gas by a factor of 3.2. Table 1.5 provides a comparison between natural gas and hydrogen as energy sources.

TABLE 1.5

Comparing Conventional Natural Gas Fuel and Hydrogen

Property	Natural gas	Hydrogen
Density at ambient (kg/m³)	0.65	0.09
Liquid density (kg/m³)	450.0	70.8
Boiling point (K)	111.2	20.3
Lower heating value (Mj/kg)	52	120
Higher heating value (Mj/kg)	47	142
Diffusion coefficient (cm²/s)	0.16	0.61
Lower flammability (vol.%)	0.07	0.04
Higher flammability (vol.%)	0.20	0.75

TABLE 1.6

Conventional Hydrocarbon Fuel Types Competing with Hydrogen as Energy Source

Dimethyl ether (DME)	C_2H_6O
Methanol	CH_3OH
Ethanol	C_2H_6O
Methoxyethane	C_3H_8O
Isobutane	C_4H_{10}
Octane	C_8H_{18}
Methylcyclohexane	C_7H_{14}

The common liquid energy carriers like methanol, propane and octane (gasoline) surpass liquid hydrogen by factors of 1.8–3.4. It is emphasized that the conversion of natural gas into hydrogen is not favorable. Although hydrogen production from a natural gas source costs less than hydrogen obtained by electrolysis, natural gas itself remains an important energy source for many applications. The overall economy of the hydrogen energy source requires 43% of higher heating value for electrolysis production and 8%–13% for compression, which increases to 60% for chemical hydride production.

Most hydrocarbon fuels contain more volumetric hydrogen (per cubic meter) than is held in the similar volume of liquefied or high-pressure compressed hydrogen. An abstract list of these chemicals is reported in Table 1.6. Some identical gases, such as ammonia, even contain near 140 kg of hydrogen per cubic meter. It is believed that the energy stored by the hydrocarbons is 2–4 times more than the energy contained in the same volume of liquid hydrogen. Other gasoline-like chemicals, such as octane, are the best hydrogen carriers with the highest energy content per volume. Therefore, the synthesis of octane from biocarbons and water is an attractive solution for an energy economy based on renewable energy sources and the recycling of carbon dioxide.

The hydrogen-powered economy is futuristic. While it has not received a considerable share in energy production, it has been utilized extensively as a chemical in many industrial and domestic processes such as the production of ammonia and methanol, refining petroleum, electronic and metallurgic processing, propellants in rockets with oxygen, gas turbine combustion fuel together with natural gas to reduce carbon dioxide emissions, and dissolved in many cosmetic products. Other advanced applications include fuel cells, a suitable component for the production of high-efficiency clean energy and reduced environmental impacts. From gas to liquid energy source, a synthetic liquid fuel must satisfy a set of requirements including being liquid under normal pressure at temperatures between $-40°C$ and $80°C$, being non-toxic, being applicable in internal combustion engines (ICEs), and being easy to synthesize and handle.

The requirements for accepting hydrogen as an energy source involve combinations of operational and thermodynamic properties. For instance, a standard of $5.5\,wt.\%$ hydrogen gravimetric capacity with a volumetric capacity of $40\,g$ of hydrogen per liter has been introduced by the US Department of Energy for light-duty vehicles. This standard guarantees a driving range of km. There is a general necessity to develop hydrogen-storage systems with refueling times of fewer than five minutes over the 1500-cycle lifetime of the system. The base, future standard, and ultimate conditions of hydrogen systems are reported in Table 1.7. According to the hydrogen-storage criteria set by the US Department of Energy such as (i) high hydrogen content; (ii) favorable thermodynamics to achieve system fill time less than $5\,min$; (iii) operation below $85°C$ and above $-40°C$ for hydrogen delivery; and (iv) cyclic reversibility (1500 cycles) at ambient temperatures.

The radar chart below (Figure 1.3) shows that despite the diversity of standard properties, most future targets are focused on hydrogen capacity with a 40%–60% improvement toward ultimate conditions.

The focus of developing technologies regarding hydrogen energy lies on hydrogen-based ICEs, Wankel engines, or nozzle jet engines, and hydrogen gas turbines alongside hydrogen fuel cells (i.e., solid oxide fuel cells). Hydrogen can be used as fuel in conventional spark-ignition engines through the controlled explosion of hydrogen–air mixtures. The reaction products of hydrogen combustion include water vapor, which further enters the biosphere. Due to the high combustion affinity of hydrogen, hydrogen-based engine efficiency is high compared to fossil-based fuels. In hydrogen combustion, the flame expands rapidly from the kernel of ignition. Besides the high-performance control of fuel induction/injection in the hydrogen engine, there is proper control over the tendency of its backfire. Hydrogen as fuel for gas turbines in critical industries such as power generation, oil and gas, process plants, and aviation was introduced as a major attempt to control greenhouse emissions (decarbonizing). The high reactivity of hydrogen gas leads to a high laminar burning velocity. Hydrogen can significantly extend the flammability limits and flame propagation of gas turbines. As a result, efficient combustion with low emission of pollutants can be guaranteed. Hydrogen fuel cells operate as electrical batteries and generate electricity through an electrochemical reaction. A typical fuel cell is composed of an anode, cathode, and an electrolyte membrane that works by passing hydrogen through the anode and oxygen through the cathode. External chemical energy reacts with hydrogen and oxygen without involving combustion. At the anode site, a catalyst

TABLE 1.7

Current, Upcoming Targets, and Future Requirements for Hydrogen as Energy Source

Storage parameter	Units	2020	2025	Ultimate
System gravimetric capacity				
Usable, specific energy from H_2 (net useful energy/maximum system mass)	kWh/kg (kg H_2/ kg system)	1.5 (0.045)	1.8 (0.055)	2.2 (0.065)
System volumetric capacity				
Usable energy density from H_2 (net useful energy/maximum system volume)	kWh/L (kg H_2/L system)	1.0 (0.030)	1.3 (0.040)	1.7 (0.050)
Storage system cost				
Storage system cost	$/kWh net ($/kg H_2)	10 (333)	9 (300)	8 (266)
Fuel cost	$/gge at pump	4	4	4
Durability/operability				
Operating ambient temperature	°C	−40/60 (sun)	−40/60 (sun)	−40/60 (sun)
Minimum/maximum delivery temperature	°C	−40/85	−40/85	−40/85
Operational cycle life (1/4 tank to full)	cycles	1,500	1,500	1,500
Minimum delivery pressure from storage system	bar (abs)	5	5	5
Maximum delivery pressure from storage system	bar (abs)	12	12	12
Onboard efficiency	%	90	90	90
"Well" to power plant efficiency	%	60	60	60
System fill time	Min	3–5	3–5	3–5
Minimum full flow rate (e.g., 1.6 g/s target for 80 kW rated fuel cell power)	(g/s)/kW	0.02	0.02	0.02
Average flow rate	(g/s)/kW	0.004	0.004	0.004
Start time to full flow (20°C)	S	5	5	5
Start time to full flow (−20°C)	S	15	15	15
Transient response at operating temperature 10%–90% and 90%–0% (based on full flow rate)	S	0.75	0.75	0.75

Note: Derived from Department of Energy technical targets for onboard hydrogen storage for light-duty vehicles.

splits the hydrogen molecules into electrons and protons. The protons pass through the porous electrolyte membrane, while the electrons are forced through an external circuit, generating an electric current and excess heat. At the cathode, the protons, electrons, and oxygen combine to produce water molecules. As there are no moving parts, fuel cells operate silently and with extremely high reliability. Solid oxide fuel cells, proton exchange membrane fuel cells, alkaline fuel cells, direct methanol fuel cells, phosphoric acid fuel cells, and molten carbonate fuel cells are the most abundantly used fuel cells [6]. A list of possible applications of hydrogen in fuel cells and their specifications is given in Table 1.8.

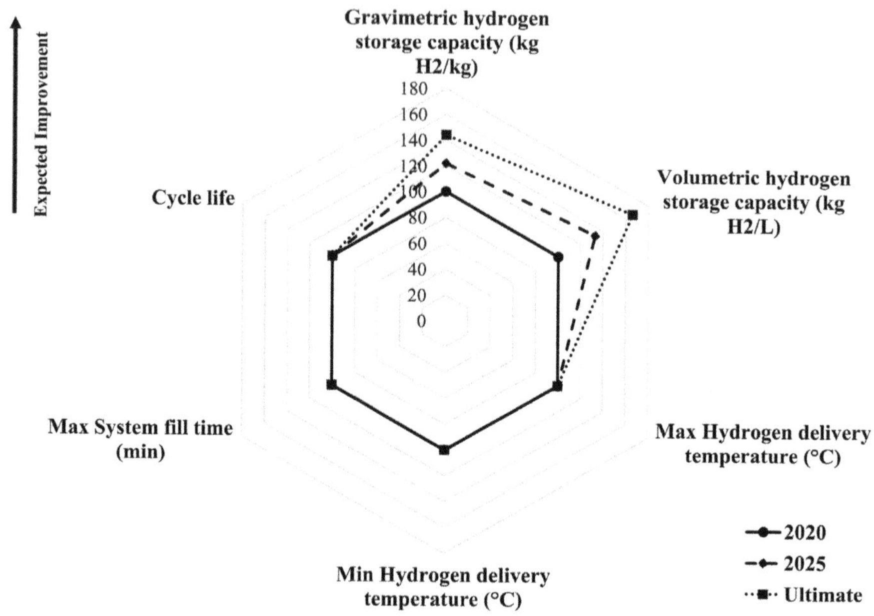

FIGURE 1.3 Hydrogen energy requirements as an energy source of the future. Created, designed, and introduced by the author.

TABLE 1.8
Fuel Cell Types as Components for Hydrogen as Energy Source

Class	Electrolyte type	Operation temperature (°C)	Efficiency (%)	Application	Disadvantage
Solid oxide fuel cells	Solid ceramic	800–1000	55	Auxiliary power units in vehicles Stationary power generation Heat engine energy recovery devices	Vulnerability to sulfur
AFC	Alkaline solution	23–70	~60	Shuttles Spacecraft	Sensitive to carbon dioxide
Direct methanol fuel cells	Polymer	60–130	40–50	Mobile electronic devices or chargers Portable power packs Electric city cars	Expensive platinum–ruthenium catalyst
Phosphoric acid fuel cells	Liquid phosphoric acid	150–200	70	Stationary applications	CO-sensitive Expensive platinum catalyst material

1.3 HYDROGEN: STORAGE AND CHALLENGES

Hydrogen consumption as an alternative energy source requires improvements in three main areas: storage, real-time production, and efficient combustion, with storage being the most challenging. Hydrogen energy storage involves storing hydrogen for later use as fuel in combustion engines or fuel cells. Hydrogen combustion can power piston engines (ICEs) and gas turbines. Electrochemical hydrogen storage in fuel cells is known for its high efficiency. On a large scale, hydrogen can be stored in large underground caverns for large-scale industrial use or in stationary/portable containers for smaller-scale applications. However, current hydrogen-storage technologies have significant drawbacks, such as complex thermal management systems, boil-off, poor efficiency, expensive catalysts, stability issues, slow response rates, high operating pressures, low energy densities, and risks of violent and uncontrolled reactions.

Physical hydrogen-storage systems are categorized based on pressure and temperature. These systems include pressure storage tanks for gaseous hydrogen, cryogenic storage tanks for liquid hydrogen, and cryogenic pressure tanks for cryo-compressed hydrogen. Technologies for physical hydrogen storage rely on pressure and temperature to increase hydrogen density. Compressed, cryo-compressed, liquid, and solid storage categories are determined by the physical state of densification. Material-based storage methods focus on how hydrogen binds to the surfaces of adsorbents and can be classified as chemical storage or absorption and physisorption. However, most adsorbent materials do not hold enough hydrogen for general applications. For example, metal hydrides add significant weight to the system, liquid hydrogen evaporates quickly, and gaseous hydrogen requires large storage volumes. Currently, the preferred hydrogen-storage solution for mobile systems is physically compressed storage with high-pressure tanks.

Linking the production of hydrogen to its final use, like any other commodity, requires processes such as transportation, storage, and transfer. Hydrogen gas should be packaged via compression or liquefaction, transported through surface containers or pipelines, stored, and then transferred to distribution outlets. Further applications require small-scale storage systems such as stationary or portable tanks and containers. Up to recently, compressed gas and liquid storage are the main methods of storage. Due to space limitations in small-scale and domestic applications and safety concerns, physically binding hydrogen-storage systems have been proposed and developed. These systems allow hydrogen gas to be physiosorbed on a high surface area adsorbent, offering a combination of safety, lightness, and affordability, along with a high hydrogen adsorption capacity, making them potential hydrogen adsorbent systems. Table 1.9 summarizes some of the most researched and reviewed hydrogen-storage methods.

Typical hydrogen supply and consumption paths as energy sources include (i) hydrogen production by electrolysis, compressing to pressures as high as 200 bar and distribution to filling stations and similar venues; (ii) hydrogen production by electrolysis, liquefying, and distribution to filling stations and similar venues; (iii) hydrogen production onsite; and (iv) hydrogen production by electrolysis to create alkali metal hydrides such as alane, magnesium hydride, alkali metal hydrides, and

TABLE 1.9

Main Storage Methods of Hydrogen as Energy Source

Method	Gravimetric density (wt.%)	Gravimetric density (MJ/L)	Temperature range (K)	Pressure range (bar)	Capacity (gH$_2$/L)
Metal borohydrides	15–19	10–17.5	130	105	70–150
Chemicals	15–16	11.5	298	10	70–150
Kubas dihydrogen complexes	10.3–10.5	23.5	293	120	100–120
Liquid organic hydrogen carriers	8.1–8.5	7	293	0	65
Metal hydrides	7.6–7.8	7.6	260–430	20	100–150
Liquid	7.2–7.5	6.4	20	0	71
Compressed	5.4–5.7	4.0–4.9	40–293	300–700	40
Metal–organic frameworks	4.5	7.5	78	20–100	55
Carbon nanostructures	2.0	5.0	298	100	~70

alanates. Alane (AlH$_3$) is a metastable hydride, stabilized by the Al affinity to combine with oxygen and form a layer of alumina ensuring chemical passivation. In bulk, alane decomposes at 100°C–150°C, and the kinetics are reasonably fast, but high hydrogen pressure is required to achieve reversibility (10 GPa, 600°C, 24 h, 10 GPa at 25°C, or 6 GPa at 300°C–380°C at very high pressures). MgH$_2$ combines properties such as availability in nature, low cost, high gravimetric (7.6 wt.%), and volumetric (110 g/L) hydrogen-storage capacities. Alkali metal hydrides such as LiH have been traditionally used for catalytic reactions but have gained attention due to their lightweight characteristics, as well as the high hydrogen gravimetric content. Nevertheless, their high thermal stability has made them less attractive in their pure form. For instance, LiH melts at 689°C and decomposes at 720°C into Li and hydrogen. Other alkali metal hydrides have unusually high decomposition temperatures due to their salt-like nature (NaH at 638°C; KH at ~400°C with K vaporizing in a hydrogen flow). Due to their high decomposition temperatures, alkali metal hydrides require kinetic and thermodynamic destabilization. Alanates, aluminum containing complex hydrides, are either octahedrally coordinated or tetrahedrally coordinated by a hydrogen atom known as hexa-hydroaluminates and tetra-hydroaluminates, respectively. The tetra-alanates (hydrido-aluminates) have a general formula M$_n$(AlH$_4$)$_n$. M is typically an alkaline ($n=1$) or alkaline earth metal ($n=2$), but M can also belong to groups III and IV in the periodic table. Examples of alanates include sodium alanates, lithium alanates, magnesium alanates, potassium alanates, calcium alanates, and mixed metal alanates. For instance, sodium aluminum hydride (NaAlH$_4$) contains 7.4 wt.% of the total amount of hydrogen but practically only 5.6 wt.% can become available [7].

Hydrogen can also be stored by physical/chemical adsorption. The mechanism involves a very tight bond between hydrogen atoms and the storage medium (metal alloys or ceramics). Heat release is essential when a hydrogen-storage system is charged. The hydrogen release at lower pressures is controlled by an influx of heat related to the hydrogen discharge rate. Storage systems based on metal hydrides store only around half wt.% of the overall estimate ($55-60 \, kg \, H_2/m^3$ compared to $70 \, kg \, H_2/m^3$ for liquid hydrogen). Physical packaging of hydrogen in hydrides can be presented for the physical (e.g., adsorption on metal hydrides) storage of hydrogen in porous matrices of special alloys, such as La-5Ni or Zr-2Cr. The energy balance situation requires measurements of the energy needed to produce and compress hydrogen in these storage systems. It is foreseeable that some of this energy input is lost in the form of waste heat. Heat must be added to release hydrogen. No additional heat is required for small discharge rates and for storage systems designed for efficient heat exchange with the environment [8]. Also, waste heat from the fuel cell may be used to heat the hydrogen-storage components. As less than two grams of hydrogen can be stored in a small $230 \, g$ metal hydride cartridge, the application of these systems for domestic and automotive usage is unforeseeable. Therefore, the energy needed to store hydrogen in physical metal hydrides is limited to the energy needed to produce and compress hydrogen to high pressures (30 bar). The energy consumption for hydrogen delivered in physical metal hydrides is much lower than that of compressed hydrogen gas delivered at 200 bar pressures. As a substitute, chemical packaging of hydrogen in hydrides is proposed. In this approach, hydrogen is stored chemically in alkali metal hydrides. The main advantage of these systems relies on various candidates from which the systems can be designed. For instance, options include the alkali group like LiH, NaH, KH, and CaH_2. Other systems include complex binary hydride compounds like $LiBH_4$, $NaBH_4$, KBH_4, and Li / $NaAlH_4$. The main disadvantage is that none of these compounds can be found in nature and should be synthesized from metals and hydrogen. The comparative relative energy needed for hydrides production is presented in Figure 1.4.

As an example, CaH_2 is produced by combining pure calcium metal with pure hydrogen at high temperature like 480°C. High energy consumption is required to extract calcium from calcium carbonate ($CaCO_3$) and hydrogen from electrolyzed water by the following endothermic processes:

$$CaCO_3 \rightarrow Ca + CO_2 + 1/2O_2 \quad E : +808 \, kJ/mol \tag{1.2}$$

$$H_2O \rightarrow H_2 + 1/2O_2 \quad E : +286 \, kJ/mol \tag{1.3}$$

The energy can be recovered when the two elements are combined at 480°C by an exothermic process:

$$Ca + H_2 \rightarrow CaH_2 \quad E : -192 \, kJ/mol \tag{1.4}$$

The combination of three equations summarizes the net reaction as follows:

$$CaCO_3 + H_2O \rightarrow CaH_2 + CO_2 + O_2 \quad E : +902 \, kJ/mol \tag{1.5}$$

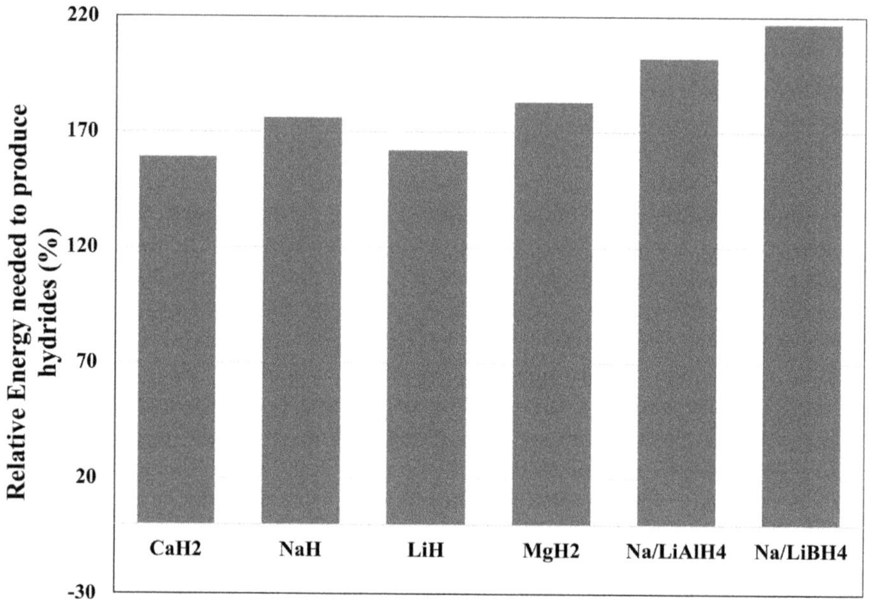

FIGURE 1.4 Hydrogen energy storage relative to energy consumption. Created, designed, and introduced by the author.

The production of other chemicals like LiH and NaH from LiCl or NaCl include following reactions:

$$LiCl + 0.5\ H_2O \rightarrow LiH + Cl + 0.25O_2 \quad E : +460\ kJ\,/\,mol \tag{1.6}$$

$$NaCl + 0.5\ H_2O \rightarrow NaH + Cl + 0.25O_2 \quad E : +500\ kJ\,/\,mol \tag{1.7}$$

Products should be cooled in a hydrogen atmosphere to room temperature, agglomerated, granulated, and packaged in airtight container systems. These hydrides react with water, releasing heat and hydrogen. The reaction with water releases twice the amount of hydrogen contained in the hydride structure. During this reaction, water is reduced while the hydride is oxidized to hydroxide. The heat generated requires cooling and is lost in most cases. Therefore, the heat energy balance should be considered during consumption.

$$LiH + H_2O \rightarrow LiOH + H_2 \quad E : -111\ kJ\,/\,mol \tag{1.8}$$

$$NaH + H_2O \rightarrow NaOH + H_2 \quad E : -85\ kJ\,/\,mol \tag{1.9}$$

$$CaH_2 + 2H_2O \rightarrow Ca(OH)_2 + 2H_2 \quad E : -224\ kJ\,/\,mol \tag{1.10}$$

An ideal storage method and medium for maximum mobility should be defined by quantifying/qualifying the characteristics of each system. Higher volumetric and gravimetric energy densities are advantageous for mobile and domestic applications. Energy densities of most liquid and gaseous fuels fluctuate due to the accessibility of complex mixtures and different blends in the supply chain. Generally, values close to 38 wt.% and 35 MJ/L have been the norm. While pure hydrogen at ambient conditions (temperature and pressure) offers excellent gravimetric energy density, its mediocre volumetric energy densities of 120 MJ/kg (100 wt.%) and 0.01 MJ/L are a concern. Another important performance property is kinetics, referring to the speed at which the system can release hydrogen upon demand and cease this release when required. Optimal rates should match transportation applications, such as acceleration and deceleration of end-users. In addition, for mobility applications, a high-power output storage system (battery) is required for peak consumption applications. Temperature-dependent hydrogen-storage methods suggest the addition of a heat management system. However, this technique is considered non-ideal and should be avoided. Operating conditions near ambient temperature throughout refueling, standby, and discharge are desirable. Operating pressure is another essential aspect to consider. High-pressure vessels as supply systems must be reinforced with high-strength materials subject to strict regulation and testing, but they negatively impact gravimetric density. System efficiency is another important thermodynamic property. Efficiency should be as high as possible to optimally utilize available renewable energy for hydrogen consumption. The efficiency has been dependent on the use of materials that require resource-intensive extraction or designs.

As the promising role of porous storage containers became clearer, similarly developed systems like metal–organic frameworks (MOFs) and Carbon nanostructures have been introduced for hydrogen storage. MOFs are a category of materials that interact efficiently with hydrogen for storage at low temperatures. Different combinations of MOF systems have been developed for applications including fuel storage, batteries, supercapacitors, photocatalysis, and phototherapy. The general hydrogen-storage condition of MOFs is attributed to hydrogen storage at low temperatures of approximately 77 K with capacities of 4.5 wt.% at 78 K and 1.0 wt.% at near ambient temperature and 20 bar. The volumetric energy density of a typical MOF can reach up to 7.2 MJ/L at 100 bar. MOF structures are composed of two main components of an inorganic metal cluster (secondary building unit) and an organic molecule called a linker; a hybrid organic–inorganic material. A schematic presentation is given in Figure 1.5 where the large open spaces between metal centers can be seen. The 3D structure provides large primary (1–1.2 nm) and secondary (0.5–0.7 nm) pores. As MOFs are porous materials composed of crystalline structures, hydrogen must inter-diffuse through crystals to be stored appropriately. The rate of hydrogen adsorption depends on the diffusivity of hydrogen into the MOF structure, permeability of MOF, and the size of the crystals (pores). The adsorption/desorption process has a fast kinetic response, in the order of seconds, which guarantees quick refueling times and cycling stability [9].

Various low-dimensional allotropes of carbon from graphite, carbon black, carbon fiber, carbon nanotubes (CNTs), and fullerene to graphene and carbon quantum dots can be applicable candidates in hydrogen-storage systems [10]. Most carbon

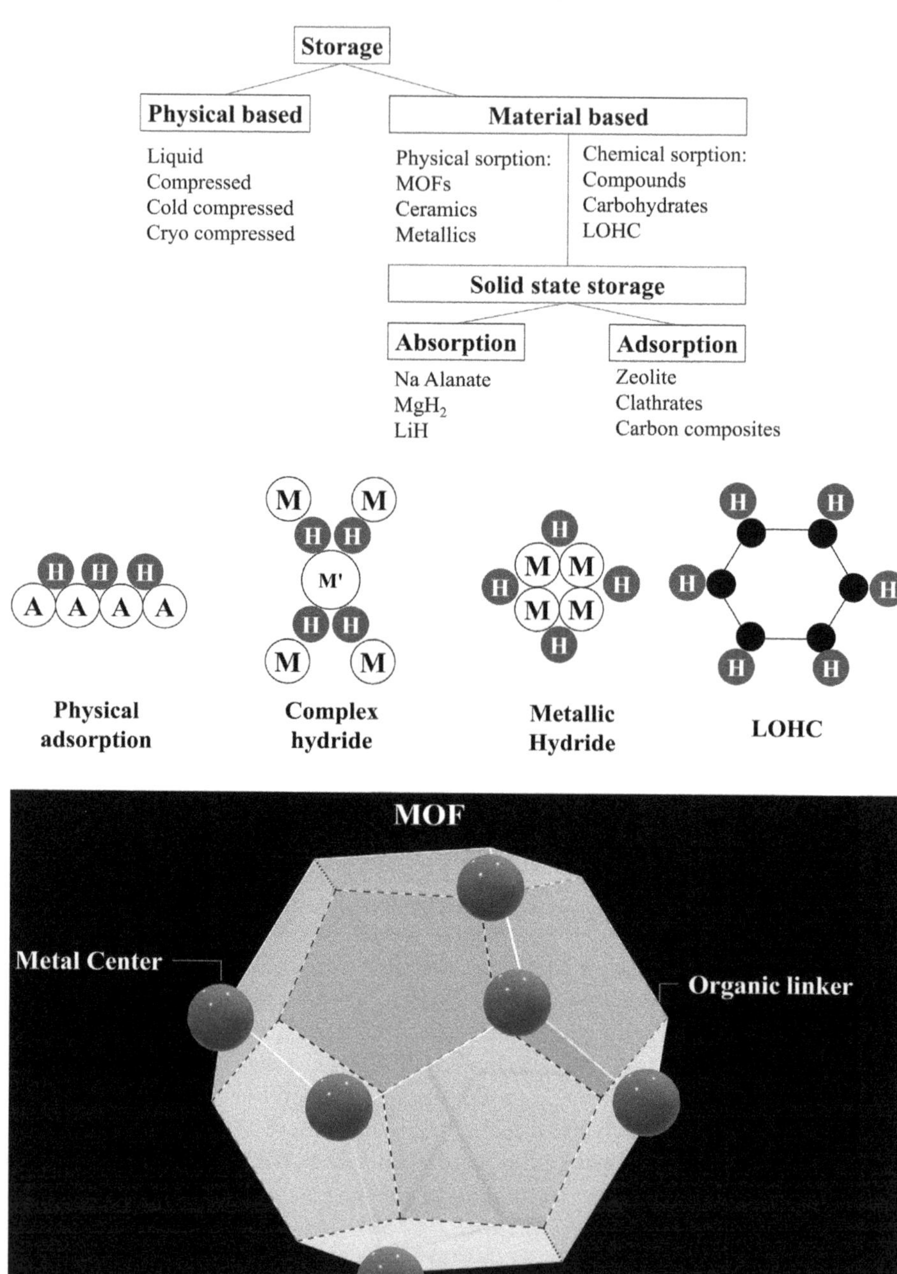

FIGURE 1.5 Schematic presentation of storage methods and materials with a structural view of process, compounds, liquid organic hydrogen carriers, and MOFs. Created, designed, and introduced by the author.

nanostructures have shown insignificant storage capacity in both doped and undoped structures. Doped CNTs with nitrogen have 2.0 wt.% at 298 K and 100 bar capacity at 298 K and 100 bar capacity, which is a very modest capacity and way below the required level. An improved structure contains significantly more hydrogen. For instance, the storage capacity of single-walled CNTs has been estimated as slightly below 10 wt.% at 298 K and 10 MPa. Other examples include a capacity of 0.298 wt.% at 20 bar when multi-walled CNTs are doped with nickel, 1.2 wt.% at 100°C in gamma ray-irradiated multi-walled CNTs, 7.3 wt.% at 20 bar, and 77 K in zeolite-templated carbon. While the level of gravimetric hydrogen-storage capacity can be interesting, the storage temperature is usually extremely low. Similar results were obtained by other carbon structures, such as graphitic with 1 wt.% hydrogen adsorption for every 500 m^2/g of surface area. Liquid organic hydrogen carriers are organic compounds that can absorb and release hydrogen through chemical reactions. Liquid organic hydrogen carriers serve as storage media for hydrogen storage. In another system, transition metal (TM) atoms form $TM - H_2$ complexes, also known as Kubas dihydrogen complexes. The $TM - H_2$ complexes are generally formed through the hybridization of the TM d-orbitals with hydrogen σ and σ* orbitals. The adsorption energy of hydrogen molecules in these complexes is usually within the range of energy required for reversible hydrogen storage at room temperature and ambient pressure (-0.4 to -0.2 eV/H_2). TM-decorated nanomaterials have attracted much attention because of their promising high capacity and reversibility as Kubas-type hydrogen-storage materials. The hydrogen-storage capacity of TM-decorated nanomaterials is as large as ~9 wt.%, which is suitable for transportation applications. In TM-decorated nanostructures, the TM atoms prefer to form clusters because of the large cohesive energy (approximately 4 eV), which leads to a significant reduction in the hydrogen-storage capacity. Figure 1.5 shows storage categories and the concept of material-based storage methods.

REFERENCES

1. Kabir MM, Akter MM, Huang Z, Tijing L, Shon HK. Hydrogen production from water industries for a circular economy. *Desalination*, 2023, 554: 116448.
2. Haoren W, Bo W, Ruize L, Xian S, Yingzhe W, Quanwen P et al. Thermodynamic analysis of the effect of initial ortho-hydrogen concentration on thermal behaviors for liquid hydrogen tanks. *International Journal of Hydrogen Energy*, 2024, 55: 243–60.
3. Zhou X, Cheng Z, Ren K, Zhai Y, Zhang T, Shen X et al. Environmental sustainability improvement in chloromethanes production based on life cycle assessment. *Sustainable Production and Consumption*, 2022, 34:105–13.
4. Anwar S, Li X. Production of hydrogen from fossil fuel: A review. *Frontiers in Energy*, 2023, 17(5): 585–610.
5. Gómez-Gualdrón DA, Wang TC, García-Holley P, Sawelewa RM, Argueta E, Snurr RQ et al. Understanding volumetric and gravimetric hydrogen adsorption trade-off in metal–organic frameworks. *ACS Applied Materials & Interfaces*, 2017, 9(39): 33419–28.
6. Das V, Padmanaban S, Venkitusamy K, Selvamuthukumaran R, Blaabjerg F, Siano P. Recent advances and challenges of fuel cell based power system architectures and control – A review. *Renewable and Sustainable Energy Reviews*, 2017, 73: 10–8.

7. Comanescu C. Recent development in nanoconfined hydrides for energy storage. *International Journal of Molecular Sciences*, 2022, 23(13): 7111.
8. Kamal MV, Ragunath S, Hema Sagar Reddy M, Radhika N, Saleh B. Recent advancements in lightweight high entropy alloys – A comprehensive review. *International Journal of Lightweight Materials and Manufacture*, 2024, 7(5): 699–720.
9. Li D, Yadav A, Zhou H, Roy K, Thanasekaran P, Lee C. Advances and applications of Metal-Organic Frameworks (MOFs) in emerging technologies: A comprehensive review. *Global Challenges*, 2024, 8(2): 2300244.
10. D'Alessandro A, Ubertini F. Advanced applications of carbon nanotubes in engineering technologies. In Abraham J, Thomas S, Kalarikkal N (Eds.), *Handbook of Carbon Nanotubes*. Cham: Springer, 2022, pp. 2001–38.

2 Ceramics
Structures, Types, and Properties

Scope

Ceramics, as non-metallic, non-organic materials, are formed through combinations of ionic and covalent bonds. The structure of ceramics can be categorized into three main groups: oxides, non-oxides, and silicates and glasses. The oxides and silicates can form crystalline and amorphous structures. Elemental components include metals binding with O, N, C, S, P, and their combinations. In this chapter, ceramic materials and structures applicable in hydrogen storage systems are introduced and reviewed.

2.1 OXIDE STRUCTURES

Oxide ceramics are those in which the non-metal oxygen is bonded with metals or metalloids. These ceramics possess different structures categorized into rock salt structure, zinc blende structure, cesium chloride structure, fluorite and antifluorite, perovskite structure, spinel structure, ilmenite structure, alumina structure, and rutile structure. These structures are representative of classes of ceramic oxides with stoichiometric or non-stoichiometric formulas of one or more cations alongside oxygen anions (ionic bonds).

In *rock salt structure*, the large anions are arranged in cubic close packing, and all the octahedral interstitial positions are filled with cations. This creates a coordination number (CN) of 6 for both ions in the ionic structure. Some of the oxides that crystallize in this structure include MgO, CaO, SrO, BaO, CdO, MnO, FeO, CoO, and NiO. For stability, the radius ratio should be between 0.732 and 0.414, and the anion and cation valencies should be the same.

The structure of *zinc blende* has tetrahedral coordination. A BeO polymorph at high temperature has this structure. The face-centered cubic (FCC) anion structure features cations filling half of the tetrahedral sites, leaving all octahedral and the remaining tetrahedral sites empty. The best representatives of this structure include ZnS, ZnO, and diamond. The radius ratio should be in the 0.225–0.414 range. The CN of Zn is 4, and that of sulfide is also 4. In the *cesium chloride structure*, the radius ratio between cesium and chloride ions requires eightfold coordination, and the bond strength is 1/8. These two criteria give rise to a simple cubic array for Cl ions. All the body center positions are filled with Cs ions. The radius ratio of Cs to Cl ions is 0.95. The *fluorite and antifluorite* structures require a 2:1 or 1:2 cation and anion ratio. The

DOI: 10.1201/9781003488828-2

best example of an oxide possessing the fluorite (CaF_2) structure is thoria (ThO_2). The charge on the Th cation is 4+, which is significantly large. This gives rise to the largest CN of 8. Therefore, the bond strength comes to 1/2. The number of valence bonds becomes four to satisfy the valency of 4 of Th to each oxygen ion. The structure is based on simple cubic packing for oxygen anions with Th cations in half the sites with eightfold coordination. The cation lattice is FCC, with all the tetrahedral interstices filled with anions. Other examples of oxides with this structure are TeO_2, ZrO_2, HfO_2, and UO_2. The similar antifluorite structure is the reverse of a fluorite structure, having a cubic close-packed array of oxygen anions with cations in the tetrahedral sites. Examples of oxides with this structure are Li_2O, Na_2O, and K_2O. The cation is coordinated with 4 ions, and anions with 8 ions. The *perovskite structure* is the first example of more complicated structures, featuring the presence of simultaneous small and large cations alongside oxygen anions. The chemical formula of perovskite is $CaTiO_3$. By applying Pauling's rule to the structure, it can be concluded that the strength of the Ti–O bond is 2/3, and that of the Ca–O bond is 1/6. Since each oxygen anion is coordinated with 2Ti and 4Ca cations, the total bond strength becomes $(4/3) + (4/6)$, which equals 2, the same as the valency of the anion oxide. The smaller cation with higher valency has a CN of 6, while the larger with a smaller valency has 12, and oxygen has 6. Other examples of compounds with a perovskite structure are $BaTiO_3$, $SrTiO_3$, $SrSnO_3$, and $CaZrO_3$. The spinel structure represents the general formula of AB_2O_4 with multi-cation oxides. An example of a ceramic possessing this structure is magnesium aluminate ($MgAl_2O_4$). The structure contains divalent A cations and trivalent B cations. The structure is cubic, having rock salt and zinc blende structures combined. A sub-cell of this structure is formed by oxygen ions in FCC packing. The four octahedral interstices and eight tetrahedral interstices are essential in this system. Out of the four octahedral sites, only two are filled in the spinel structure, and out of the eight tetrahedral sites, only one is filled. Eight elementary cells are arranged to form a unit cell. Spinel is categorized into two main types: normal and inverse. In normal spinel, the A cation ions occupy tetrahedral sites and B cations occupy octahedral sites. Examples of normal spinel are $ZnFe_2O_4$, $CdFe_2O_4$, $MgAl_2O_4$, $CoAl_2O_4$, $MnAl_2O_4$, and $ZnAl_2O_4$. In inverse spinel structures, A cations and half of the B cations are located in octahedral sites, while the other half are positioned in tetrahedral sites. This arrangement is commonly represented as $B(AB)O_4$. These reverse-structured compounds are more common than normal spinel. Some examples of R-spinel include $Fe(MgFe)O_4$, $Fe^{3+}(Fe^{2+}Fe^{3+})O_4$, and $Fe(NiFe)O_4$, especially among magnetic ferrites. The *alumina* structure (Al_2O_3) is a common structural arrangement shared by many oxides in the transition metal system. In alumina structures, the preferred CN for Al ions is 6, and the bond strength is equal to 0.5. In the structure, there are four Al ions adjacent to each O ion to create a total bond strength equal to the valency of the oxide anion. This structure is based on a unit cell of hexagonally closed packed O ions, with Al ions filling two-thirds of the octahedral sites. The *ilmenite* structure type is based on iron titanate ($FeTiO_3$). The $FeTiO_3$ crystal structure is an ordered derivative of the alumina structure. In the ilmenite structure, the two Al ions are replaced by a divalent Fe ion and a tetravalent Ti ion in the formula. The structure is based on hexagonally closed packed of O ions, with cations filling two-thirds of the octahedral sites. Half the cation sites are occupied by Fe, and the rest are filled by Ti ions. Other examples of ceramics with ilmenite structure are $MgTiO_3$, $NiTiO_3$, $CoTiO_3$, and $MnTiO_3$. The rutile structure is based on titanium dioxide (TiO_2), where the CN for quadrivalent Ti is 6, and therefore the bond strength is 2/3. This arrangement creates threefold coordination for O ions. Cations fill half the available octahedral sites, and the closer packing of O ions around cation sites causes distortion of the anion FCC base lattice. Other examples of compounds with rutile structures are GeO_2, PbO_2, SnO_2, and MnO_2. The schematic presentation of structures is shown in Figure 2.1.

Silicates are complex oxide ceramics that consist of combinations of two or more oxides based on silica building blocks (ionic–covalent bonds). In silicates, two types of oxygens are formed: bridging oxygens (BOs) and non-BOs (NBOs). In BOs, oxygen atoms create bonds with two silicon (Si) atoms, whereas in NBOs, they are bonded to only one Si atom. NBOs are formed by the addition of either alkali or alkaline earth metal oxides to silica, where O− denotes an NBO. NBOs carry a negative charge, while overall neutrality is maintained by cations being adjacent to the NBOs. The number of NBOs is proportional to the number of moles of alkali or alkaline earth metal oxides added. The addition of these oxides to silica increases the overall O/Si ratio of the silicate. Increasing the number of NBOs results in the progressive breakdown of the silicate structure into smaller units. The number of NBOs per tetrahedron is a critical parameter that determines the structure of silicates, which is influenced by the O/Si ratio. The fundamental building block of silicates is SiO_4^{4-} tetrahedron. The Si–O bond is partly covalent, and the tetrahedron fulfills the bonding requirements of covalent directionality and the relative size ratio for ionic structures.

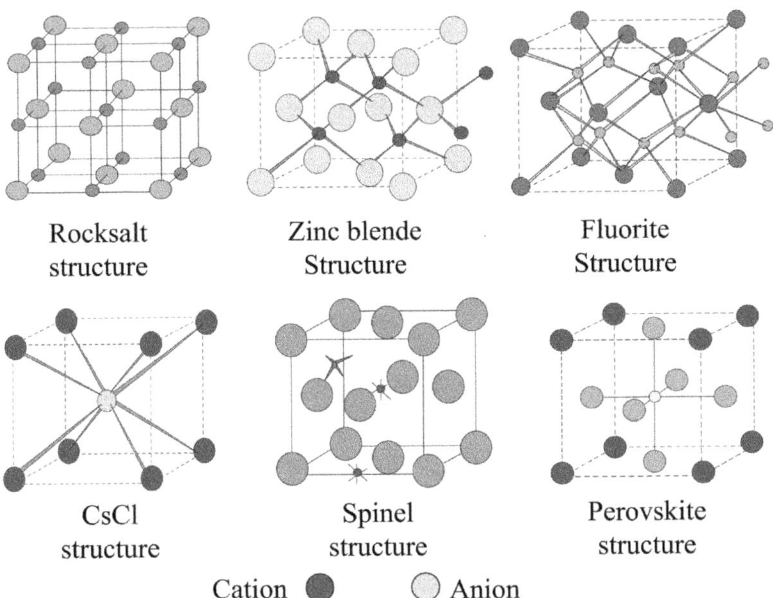

FIGURE 2.1 Schematic presentation of ionic and semi-ionic ceramic structures. Created, designed, and introduced by the author.

Due to the high charge on the Si ion, the tetrahedral units are rarely joined edge to edge and never face to face; instead, they almost always share corners. As one moves from corner to edge to face sharing, the cation separation distance decreases. This, in turn, results in cation–cation repulsions and a decrease in structural stability. The distances of cation–cation repulsion increase from left to right, which tends to destabilize the structure. The *silica* structure is created exclusively by the SiO_4^{4-} tetrahedron building block. When this block is connected to neighboring tetrahedra, the four negative charges on each block are balanced, and the structure becomes neutral. This gives rise to an O/Si ratio of 2 and the chemical formula of silica, where each O is linked to two Si atoms and each Si is linked to four O atoms. Since the tetrahedron is a three-dimensional structure, the network formed by the combination of tetrahedra also becomes a three-dimensional network. Examples of this kind of structure include the different allotropes of silica such as quartz, tridymite, and cristobalite. In the absence of long-range order, the resulting solid is known as amorphous silica or fused quartz.

Sheet silicates are created when three out of four oxygens are shared, resulting in an O/Si ratio of 2.5. Examples of this structure include clays such as kaolinite($Al_2(OH)_4(Si_2O_5)$), talc ($Mg_3(OH)_2(Si_2O_5)_2$), and mica ($KAl_2(OH)_2(AlSi_3O_{10})$). The $(Si_2O_5)^{2-}$ sheets are held together by positively charged ions, which explains the water absorption by clays. The polar water molecule is easily absorbed between the tops of positive sheets and the bottoms of silicate sheets. Al ions substitute for one-fourth of Si atoms in the sheets, which requires an alkali ion, such as K, to remain electrically neutral. Alkali ions fit into the holes of silicate sheets and bond the sheets together with ionic bonds, which are stronger than the base bonds. In Chain Silicates, an O/Si ratio of 3 creates infinite chains or ring structures (i.e., asbestos). Chains are held together by weak electrostatic forces. Island Silicates are formed when the O/Si ratio is 4, resulting in isolated SiO_4^{4-} tetrahedra connected by positive ions. The resulting structure is known as an island silicate. In garnets $(Mg,Fe^{2+}, Mn, Ca)_3(Cr,Al,Fe^{3+})_2(SiO_4)_3$ and olivines $(Mg,Fe^{2+})_2(SiO_4)$, the SiO_4^{4-} becomes the anion and is bonded ionically by the metallic cations. Aluminosilicates are formed by the combination of alumina and silica. Similar to the mica structure, Al ions substitute for Si ions; the resulting negative charge must be compensated for by an additional cation. In clays, Al ions occupy the vacancies in the silicate network. When Al substitutes for Si in the network, the appropriate ratio for determining the structure is the O/(Al + Si) ratio. For albite ($NaAlSi_3O_8$), anorthite ($CaAl_2Si_2O_8$), eucryptite ($LiAlSiO_4$), orthoclase ($KAlSi_3O_8$), and spodumene ($LiAlSi_2O_6$), the ratio becomes 2. The structure is three-dimensional and corresponds to three-dimensional tetrahedral units. As the bonds are both covalent and ionic primary bonds, the melting points of these aluminosilicates are significantly high. In these structures, bonding can be mixed within the silicate network. Si–O–Si bonds differ from the bonds holding the units together, which can be ionic or weak secondary bonds. The schematic representation of some oxides/silicates is shown in Figure 2.2.

Glasses are inorganic non-crystalline solids, and their structures are on the scale of 2–10 angstroms for local atomic arrangements, while the scale of 30 to a few thousand angstroms occurs as a sub-microstructure, and the scale of microns to millimeters or

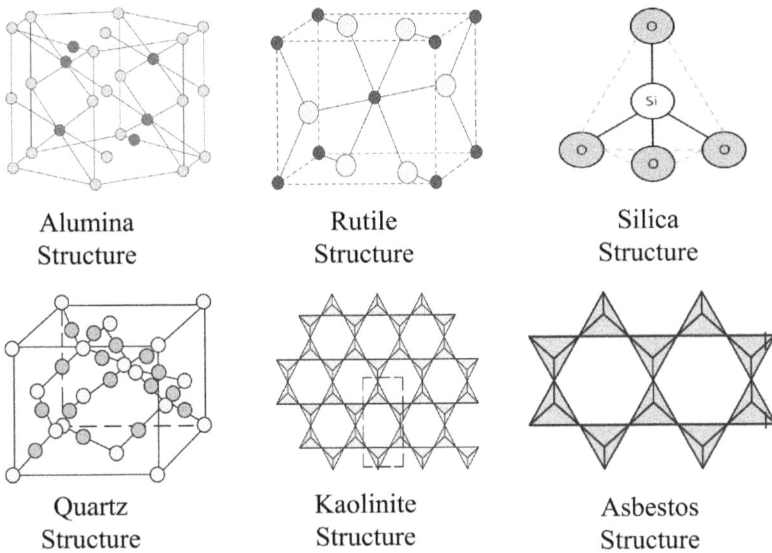

FIGURE 2.2 Schematic presentation of ionic–covalent ceramic and silicate structures. Created, designed, and introduced by the author.

more is considered the microstructure and macrostructure. On the atomic scale, the distinguishing structural characteristic of glasses is the absence of atomic periodicity or long-range order. A typical glass structure follows four rules for the formation of an oxide glass: (i) Each oxygen ion should be linked to no more than two cations, (ii) the CN of oxygen ions around the central cation must be small, four or less, (iii) oxygen polyhedra share corners, not edges or faces, and (iv) at least three corners of each polyhedron should be shared. Silica and borate are the primary glass base structures. The addition of alkali and alkaline earth oxides to silica and boron trioxide results in the formation of SiO_4 and BO_4 tetrahedra. Similarly, germanate and phosphate glasses are based on glassy germanium (IV) oxide forming GeO_4 tetrahedra, with a mean Ge–O–Ge bond angle of about 138°. In contrast to fused silica, the distribution of inter-tetrahedral angles (Ge–O–Ge) for germanium is quite sharp. As with silicate and most germanate glasses, phosphate glasses are composed of oxygen tetrahedra; however, unlike silicate and germanate glasses, a PO_4 tetrahedron can be bonded to at most three other similar tetrahedra. The most familiar structural units in phosphate glasses are rings or chains of PO_4 tetrahedra. A schematic presentation of the molecular structure of crystalline ceramic and glass in comparison with liquid is shown in Figure 2.3.

2.2 NON-OXIDE STRUCTURES

Non-oxide ceramics include carbides, nitrides, borides, silicides, and others. The applications of non-oxides range from abrasives (B_4C, BN) and cutting tools (WC), to seals and wear parts (TiB_2), electrodes for metal melts, refractory materials in foundries, electrical devices, nozzles, armor (ZrB_2), and heating elements ($MoSi_2$).

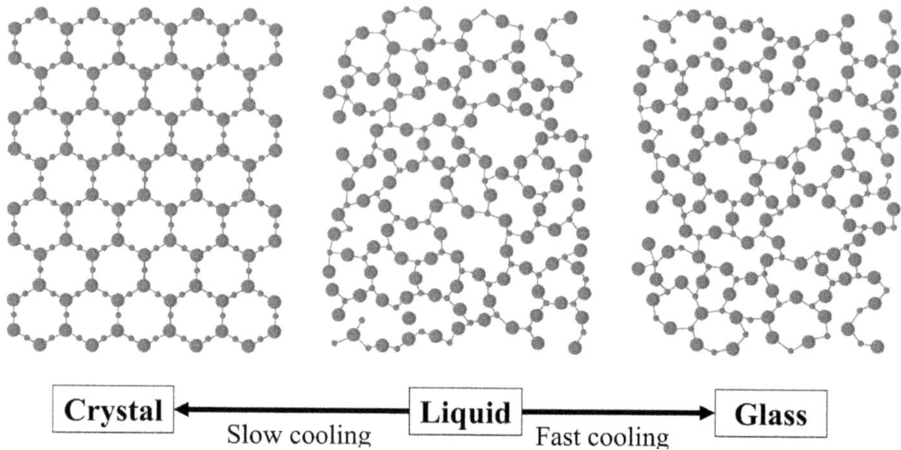

FIGURE 2.3 From liquid (melt) to crystalline or glass structure. A molecular comparison. Created, designed, and introduced by the author.

TABLE 2.1

Non-oxides Ceramics and Examples of Their Applications

Non-oxide Ceramic	Application	Specification
$SiC–Si_3N_4$	Thermal insulation of high-temperature components	Wear resistance Heat and electrical insulation Low density Resistance to corrosion, High temperature strength
SiC–C (graphite)	Chemical industry components	Resistance to corrosion Wear resistance
Si_3N_4, SiC, C, BN, $MoSi_2$, ZrB_2	Heat exchange Crucibles	Resistance to corrosion Thermal and electrical insulation High temperature strength
Si_3N_4, SiC, B_4C, TiC, TiN, BNdiamond	Cutting and grinding tools	Resistance to corrosion Wear resistance
AlN	Integrated circuits	Electrical insulation Heat conductivity

The most important structural non-oxide ceramics are silicon carbide SiC, silicon nitride Si_3N_4, and the so-called sialons – nitride-based ceramics with varying oxide contents (Table 2.1). Non-oxides require high-temperature processing in reducing (hydrogen) or inert (nitrogen and argon) atmospheres to prevent oxidation. Furthermore, their predominantly strong covalent (i.e., directional) atomic bonds limit atomic migration (diffusion) and restrict solid-state sintering below the decomposition temperature (~2500°C for SiC and 1900°C for Si_3N_4). Liquid-phase sintering or reaction-bonding techniques are necessary for densification.

In *silicon nitride* polymorphs and sialon structures, the atomic bonding of Si_3N_4 is about 70% covalent (with building blocks in the form of SiN_4 tetrahedra). Si_3N_4 has two polymorphs: α (low temperature) and β (high temperature), both of which are hexagonal, with α- Si_3N_4 being harder. The equilibrium α-β transition occurs at 1400°C, although in practice, this often shifts to much higher temperatures. Typical properties of Si_3N_4 include a density of 3.2 g/cm³, hardness between 14 and 18 GPa, a fracture toughness of $2-8$ MPa·m$^{1/2}$, Young's modulus of 320 GPa, bending strength between 400 and 1000 MPa, a thermal expansion coefficient of $2-3\times10^{-6}$ K^{-1}, and thermal conductivity on the order of 50 W/mK; this combination results in a low thermal expansion coefficient and high thermal conductivity, along with excellent thermal shock resistance.

In *sialons*, Al^{3+} is substituted on some of the Si^{4+} sites and O^{2-} on some of the N^{3-} sites in Si_3N_4 crystal structures. In α-sialons, more Si is substituted by Al than N by O. As a result, charge-compensating cations are required to achieve charge balance (e.g., Mg, Ca, Y, and Ce). In β-sialons, equal numbers of Al and O are substituted for the Si and N sites, so no additional cations are necessary for charge balance. The general sialon chemical formula is $Si_{6-z}Al_zO_zN_{8-z}$ (where $0<z<4.2$). Since the unit cell dimensions of β-sialons are functions of z, the overall formula includes (in nm) $a=0.7603+z\cdot0.0030$ and $c=0.2907+z\cdot0.0026$. The exact lattice parameters are obtained by electron or X-ray diffraction to determine the value of z for a given material. By varying the overall composition in the Si-Al-O-N system (with charge-compensating cation), it is possible to create compositions of α and β sialons where properties such as hardness and fracture toughness can be controlled. α-sialons (typically equiaxed) have a higher hardness, while β-sialons (typically high aspect ratio needles) have higher strength and toughness. β-Sialons have similar physical properties to Si_3N_4, but exhibit some improved chemical stability. Typical properties of sialons are: density 3.25 g/cm³, hardness 15–24 GPa, fracture toughness $4-8$ MPa·m$^{1/2}$, bending strength up to 1000 MPa, thermal expansion coefficient 3×10^{-6} K^{-1}, and thermal conductivity up to 10–50 W/mK. The strength of sialons is high at room temperature but decreases at higher temperatures. Because of their high hardness and wear resistance, SiAlON and sialons are used as abrasives, grinding media, cutting tools, armoring, and nozzles. They also demonstrate resistance to high temperatures and chemical attack, making them suitable for use in burners, welding nozzles, heat exchangers, and as kiln furniture. Combined with low density and suitable electrical properties, they are applicable for engine parts (valves, turbocharger rotors, gas turbines, catalyst carriers, spark plugs). Finally, due to their good thermal shock resistance, they can be used for crucibles, thermocouple shielding tubes, etc. Sialons have the additional advantages of higher toughness, improved chemical stability, and sinterability. They are mainly used in applications requiring high wear resistance.

Silicon carbide (SiC) is the most widely used non-oxide ceramic material. It was originally synthesized by heating a mixture of clay and carbon at 1600°C in a carbon arc (carborundum for abrasive and refractory grade SiC). Most SiC (α-SiC) is synthesized via a similar process (carbothermal reaction of quartz). For SiC (β-SiC) platelets or whiskers, other processes are used, such as pyrolysis of rice hulls, spinning from organosilicon polymer precursors, chemical vapor deposition, and the so-called vapor–liquid–solid process, which uses liquid catalysts. Due to the difference in electronegativity between carbon and silicon, the structure is predominantly covalent (88%) with strong, directional bonds, which are responsible for the high decomposition temperature and hardness of SiC. Silicon carbide crystallizes in two general crystalline forms: cubic β-SiC (low temperature) and α-SiC (high temperature). Due to one-dimensional polymorphism called polytypism (i.e., various stacking sequences in the c-direction), α can take on a special type and can be cubic (zincblende structure) or hexagonal (wurtzite structure). Usually, α-SiC is obtained by carbothermal reduction of quartz, while β-SiC is formed by direct reaction of silicon and carbon (self-sustaining reaction), vapor phase routes, and using plasma or laser heating. SiC powders always have an oxide coating (silica surface layer), and particle size control is necessary because ultrafine powders have high oxygen impurity. Graphitic carbon and silicon impurities can influence processing and high-temperature behavior. A sufficient amount of carbon must be present to react with the surface oxide and allow good sinterability. Fine-grained microstructures with isometric grains are desired in SiC to ensure the best mechanical properties. Typical properties of SiC include density 3.2 g/cm^3 , hardness 21–25 GPa, fracture toughness $3-6$ MPa$\cdot$m$^{1/2}$, Young's modulus 420 GPa, bending strength 450–650 MPa, thermal expansion coefficient $4-5\times10^{-6}$ K^{-1}, and thermal conductivity on the order of 50 W/mK. The creep resistance of SiC is higher than that of silicon nitride. The main application of low-grade SiC is as abrasives and in grinding wheels. SiC is used as a refractory material (due to its high thermal conductivity, high decomposition temperature, chemical inertness, and low wettability by molten metals and slags), heating element (due to its appropriate electrical properties), high-temperature semiconductor, and structural material in contact with corrosive liquids such as concentrated HF and NaOH (due to its excellent corrosion resistance). It is also used in heat engine components due to its excellent wear resistance, as the low coefficient of friction reduces wear in contact with itself or other materials. Schematic presentations for typical silicon non-oxide microstructures are shown in Figure 2.4.

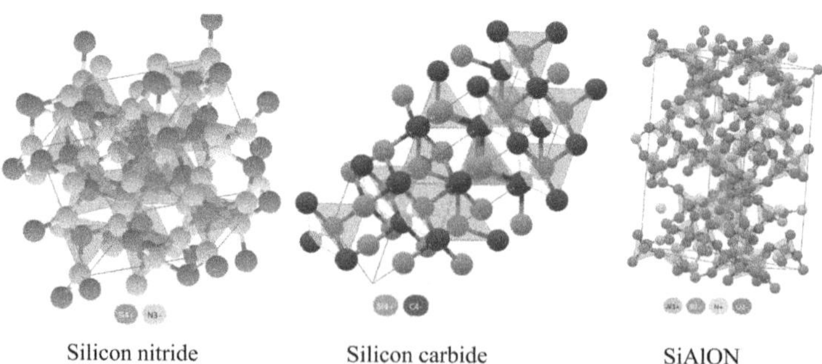

FIGURE 2.4 Microstructure of typical silicon non-oxides. Created, designed, and introduced by the author.

TABLE 2.2

Comparison between Carbides and Nitrides Based on Decomposition Temperature and Density

Carbide	Melting temperature (°C)	Density (g/cm^3)	Nitride	Melting temperature (°C)	Density (g/cm^3)
HfC	~3890	<12.2	HfN	~3310	14.1
TaC	~3880	14.5	TaN	~3100	~14.0
ZrC	~3530	<6.8	ZrN	~2980	7.4
TiC	~3250	<4.8	TiN	~2950	4.0
ThC	~2620	10.7	ThN	~2630	~11.5
Al_4C_3	~2800	3.1	AlN	~2400	3.1
B_4C	~2700	~2.4	BN	~3000	>3.3
MoC	~2690	8.5	VN	~2030	6.0
WC	~2850	~15.5	Si_3N_4	~1900	~3.2
SiC	~2700	~3.2			

TABLE 2.3

Comparison between Borides and Silicides Density and Crystal Structures

Boride	Crystal structure	Density (g/cm^3)	Silicide	Crystal structure	Density (g/cm^3)
HfB_2	Hexagonal, hP3	10.5	HfSi	Orthorhombic Pnma	1.9
ZrB_2	Hexagonal, hP3	6.1	$ZrSi_2$	Orthorhombic	4.9
TaB_2	Hexagonal, hP3	11.2	$TaSi_2$	Orthorhombic	9.1
TiB_2	Hexagonal, hP3	4.5	$MoSi_2$	Tetragonal	6.3

Other carbides and nitrides are similarly formed from three main families: alkaline metal or alkaline earth carbides; interstitial carbides with transition metals (e.g., Ti, Zr, Hf, V, Nb, and Ta); and covalent macromolecular carbides with B, Al, and Si. Only compounds from the last two families are used for preparing ceramics. Only interstitial nitrides (TiN, ZrN, HfN, TaN) and macromolecular covalent nitrides (Si3N4, AlN, BN) have been significantly developed. A comparison between carbides and nitrides is listed in Table 2.2. The data clearly show the high-temperature refractory properties of non-oxides. Similar characteristics can be found in borides and silicides. Check Table 2.3 for functional information on silicides and borides, including crystal structure. See Figure 2.5 for some of the well-known microstructures.

2.3 CARBON-BASED CERAMICS

The versatile arrangements of C atoms and the formation of different carbon allotropes and phases have led to a wide range of carbon-based materials and structures due to four valence electrons (2s and 2p) and their participation in the formation of single, double, and triple bonds. Carbon atoms can react with more electronegative and

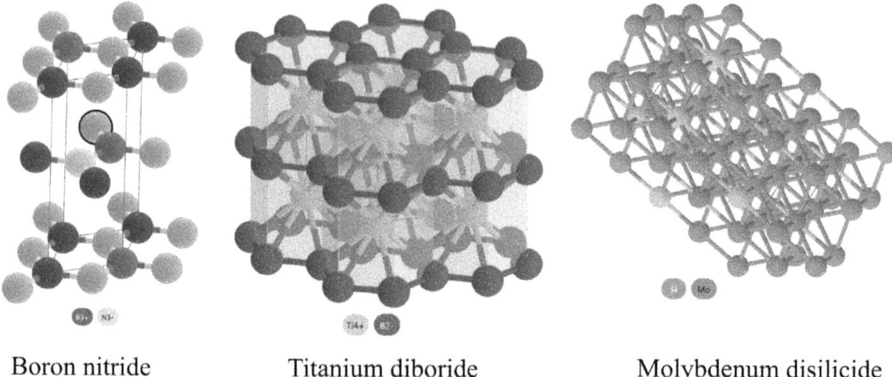

FIGURE 2.5 Microstructure of typical nitride, boride, and silicide. Created, designed, and introduced by the author.

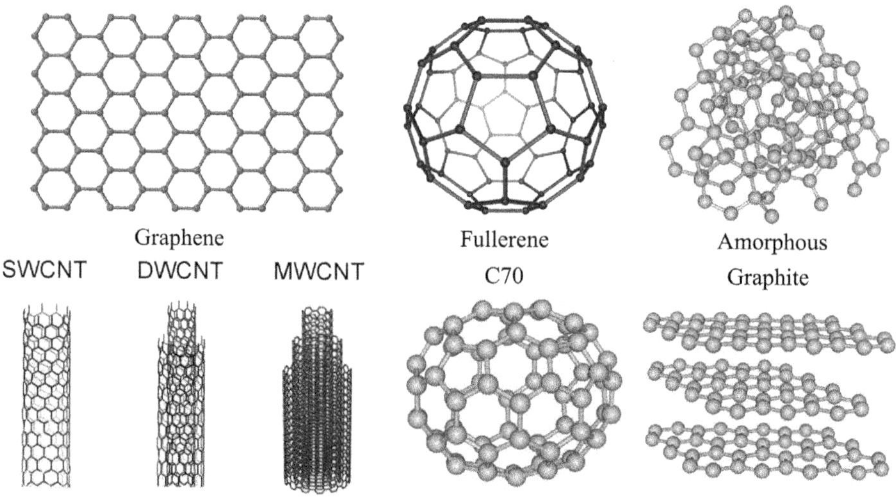

FIGURE 2.6 Schematic presentation of carbon allotropes. Created, designed, and introduced by the author.

electropositive elements to form stable double and higher-order atomic compounds. Enabled by three hybridization forms (sp, sp^2, and sp^3), many different amorphous and crystalline structures, as well as various forms and dimensions, can be created from carbon, including 0D fullerene molecules; 1D single- and multi-walled carbon nanotubes (CNTs) and nanofibers nanofibers; 2D carbon nanohorns and graphene; and 3D foams, aerogels, and hydrogels. Figure 2.6 presents the main categories of carbon allotropes. Graphite, as a natural material formed in crystalline, amorphous, and lumpy forms, is the most stable form of carbon. It consists of layers of sp^2 hybridized carbon atoms and a delocalized cloud of π-electrons. These individual layers are called graphene sheets or graphene layers, which are bound together only by weak van der

Waals' forces. *Diamond* is an allotrope of carbon, formed by a network of sp^3 hybridized carbon atoms arranged in a cubic-centered lattice. Carbon atoms are bonded to and surrounded by four other atoms, forming a tetrahedron. Diamond theoretically exists in two modifications of hexagonal and cubic symmetries. Besides its natural rarity, synthesized diamond is being industrially used in nanostructured diamond or ultrafine nanostructured diamond, which consist of nanocrystalline diamond grains ~5 nm in size, embedded in an amorphous carbon matrix. *Graphene*, as a 2D allotrope, is a monolayer of sp^2 hybridized carbon atoms with a hexagonal lattice (one layer of graphite). Carbon atoms have three σ-bonds, and a p-bond outside the plane that can bind to other atoms. Graphene can also exist in a non-flat 2D form with ripples, wrinkles, folds, and creases. Other graphene-based materials, produced through chemical and physical modifications, include multilayer graphene, graphene oxide (GO), reduced graphene oxide (rGO), and carbon nanotubes and nanohorns. Processing graphene is limited by its poor solubility and dispersibility in liquids. *Carbon quantum dots* (CQDs) and *graphene quantum dots* (GQDs) usually contain amorphous or nanocrystalline nuclei with sp^2 hybridized carbon. CQDs are nano-dimensional objects with a size below 20 nm, while GQDs usually have a diameter of about 10 nm. The special characteristics of CQDs and GQDs are related to the comparable size of materials in comparison to the wavelength of the electron. *Fullerenes* (Buckminsterfullerene) are a carbon allotropic structure that consists of an even number of sp^2 hybridized carbon atoms. Carbon atoms are connected in 12 pentagonal and m hexagonal rings, where $m = (n-20)/2$, where n denotes the total number of carbon atoms in the molecule. C60 (12 pentagons and 20 hexagons) is an example of the smallest fullerene. Due to the curvature in the structure, fullerenes experience pyramidalization through a change in hybridization from pure sp^2 to an intermediate state between sp^2 and sp^3 hybridization. *CNTs* are considered to have a simple atomic configuration of graphene sheets rolled into cylindrical shapes. Depending on the number of graphene layers, they are divided into single-walled CNTs, which contain only one layer of graphene, and multi-walled CNTs, which of layers spaced 3.4 Å apart with a diameter in the 10–20 nm range. *Carbon nanofibers* (CNFs) are formed by rolling graphene sheets at a certain angle (α), creating a stack of nanocones. CNFs differ from carbon nanotubes by their curving angle, which is equal to zero for CNTs. The size of the CNF diameter varies from 3.5 nm to several hundred nanometers, with a length of up to several micrometers. Substructures of CNFs include hollow and filled cores, stacked nanocones, partitioned stacked nanocones, and partitioned nanotubes. Both structures have a very low density, and their diameter is significantly smaller than their length.

The application of CNTs is limited due to difficult dispersion and poor interfacial interaction. The carbon nanostructures include *carbon nano-onions*, a multilayer fullerene structure, a combination of fullerenes and multi-walled nanotubes, and fullerene cages. Multilayer graphitic structures can be categorized as bulbous graphite nanoparticles (which have a faceted shape and a significantly larger internal cavity) and real carbon onions (which have a concentric structure of spherical shells). These structures contain hexagonal and pentagonal rings in which carbon atoms are located at the vertices and form two single bonds and one double bond with adjacent carbon atoms, along with delocalized π-electrons. See Figure 2.7 for a representation of core–shell and carbon nano-onion allotropes.

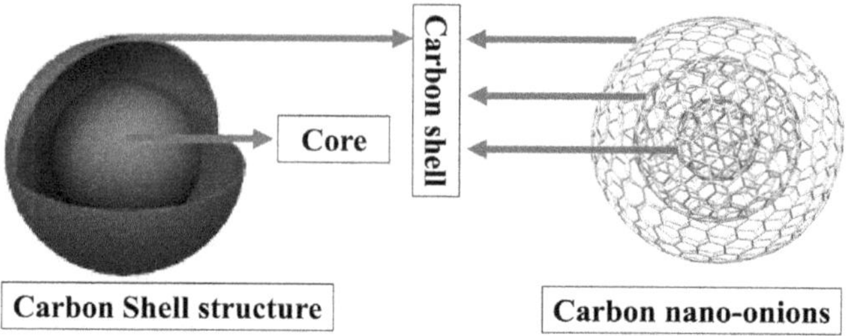

FIGURE 2.7 Yolk–shell carbon microspheres and carbon nano-onion allotropes. Created, designed, and introduced by the author.

Carbon nanoparticles, which are combinations of activated carbon, charcoal, and carbon black, are formed with a high specific surface area due to their particle structure. CNPs are products of incomplete combustion, and carbon black is obtained through thermal decomposition. *Carbon aerogel* and *aeroraphite* are low-density, large-surface-area, and controllable stable structures of carbon materials formed as 3D nano-grids with an open porous structure that facilitates the penetration of molecules and ions into interior spaces. These structural features have made them preferable as absorbents and catalysts for hydrogen fuel storage. The microstructure consists of interconnected solid phase particles and carbon grains. There is a network of interwoven graphite fibers with a width of several nanometers (nm). The main methods of aerogel preparation include (i) direct carbonization of aerogels, (ii) self-assembly of nanocarbon building blocks, and (iii) chemical vapor deposition. Similarly, graphene-based aerogels have shown high adsorption potential and excellent recyclability.

2.4   TEXTURAL CERAMICS

Textural ceramics are a sub-category of bulk ceramics with properties that depend on both topology and material, compared to monolithic/composite bulks, where properties are derived from densification state and material. These ceramics are rapidly gaining importance as adsorbents, sorbents, filters, catalyst supports, membranes, and scaffolds for cell growth. The main design and development principles that influence cellular properties are (i) the properties of the solid precursor ceramics, (ii) the topology (connectivity) and shape of building blocks, and (iii) the relative density ($\tilde{\rho} / \rho_s$) of the ceramics, where $\tilde{\rho}$ is the density of the textural ceramic and ρ_s is that of the solid precursor ceramic. Textural ceramics, also known as cellular or "lattice" ceramics, are composed of a lattice, a connected network of structural components. In textural ceramics, a lattice truss or space frame consists of arrays of a three-dimensional arrangement of units, which are pin-jointed or rigidly bonded at their connections. Textural ceramics are designed to be load-bearing, highly spacious, and porous, resulting in light blocks with a high strength-to-density ratio. Similar to a crystal lattice, textural bulks can be characterized by a typical cell with certain symmetry elements, and in some cases, with translational symmetry. While the scale of a typical crystal lattice is at the atomic level, the unit cell of textural ceramic materials is on the scale of millimeters or micrometers. These structures could include honeycombs, ceramic foams, hollow spheres, carbon foams, glass foams, and three-dimensional periodic structures.

Honeycomb ceramics are porous ceramics created from a paste of a viscoplastic, formable body of dispersed particles in a polymer solution. This paste process yields a porous, particulate honeycomb with a microstructure that can be processed to change its composition, porosity, and connectivity. The honeycomb shape can be controlled by extrusion conditions, enabling a wide range of honeycomb channel diameters from 1 cm to less than 0.5 mm. The honeycomb extrusion process is a mechanical method to create linear porosity. A porous ceramic is a composite of nonsolid (gas and/or liquid) and ceramic materials. It can have the surface area of ceramic powder in the shape of an object. If the nonsolid phase is open and continuous, the entire volume of the object is readily accessible for reaction. The distribution of the nonsolid in a solid includes random (dispersed bubbles and percolating paths) and nonrandom (honeycombs and woven structures) arrangements, which create high surface area with optimal permeability. The honeycomb ceramics have two types of porosity: geometric (channels) and microstructural (pores and microcracks). The cross-sectional geometry of the honeycomb ceramics includes a tessellating pattern where the channels are combinations of triangular, square, hexagonal, round, or other shapes. Designs can include gradients in pitch and web thickness in the radial and axial directions made from alumina, carbon, cordierite, fused silica, mullite, potassium borosilicate glass, silicon carbide, and similar materials. Ceramic foams, which are highly porous ceramic materials, offer characteristics such as high surface area, high permeability, low density, low specific heat, and high thermal insulation. These characteristics are essential for applications such as gas adsorption, catalyst supports, filters for molten metals, hot gas filters, ion exchange, refractory linings for furnaces, thermal protection systems, heat exchangers, and porous implants in biomaterials. The high porosity property is related to cell size, morphology, and degree of interconnectedness. Open-cell and interconnected ceramics are suitable for uses involving gas transport, such as physical adsorbents. Most ceramic filters are produced by replicating polymer foams through applying a ceramic slurry (with solid content in the range of 50–70 wt.%, exhibiting thixotropic behavior, where the viscosity decreases with time at a fixed rate of shear), which is dried in place prior to the polymer template burn-out and the subsequent sintering of the ceramic (with density and permeability being important factors). Different polymers are used for the precursor foam, including polyurethane, poly(vinyl chloride), polystyrene, and cellulose. This method creates very open, reticulated foams with hollow structures (97% void volume with pore sizes between 5 and 65 ppi (2–25 pores per cm) and densities ranging between 5% and 30% of theoretical density) during the burn-out stage. Another method involves foaming a ceramic slurry through mechanical agitation or in situ evolution of gases, resulting in a wide range of open-pored structures with lower porosity than replicated foams. The other approach incorporates sacrificial additives into the ceramic structure to create open or closed-pored foams. In situ gas evolution enables the foaming of ceramics by the presence of a foaming agent that decomposes due to heat or a chemical reaction, generating gas within the ceramic slurry. These ceramic foams have an open-celled structure with pore size related to the reticulation factor, as shown in Figure 2.8 schematically.

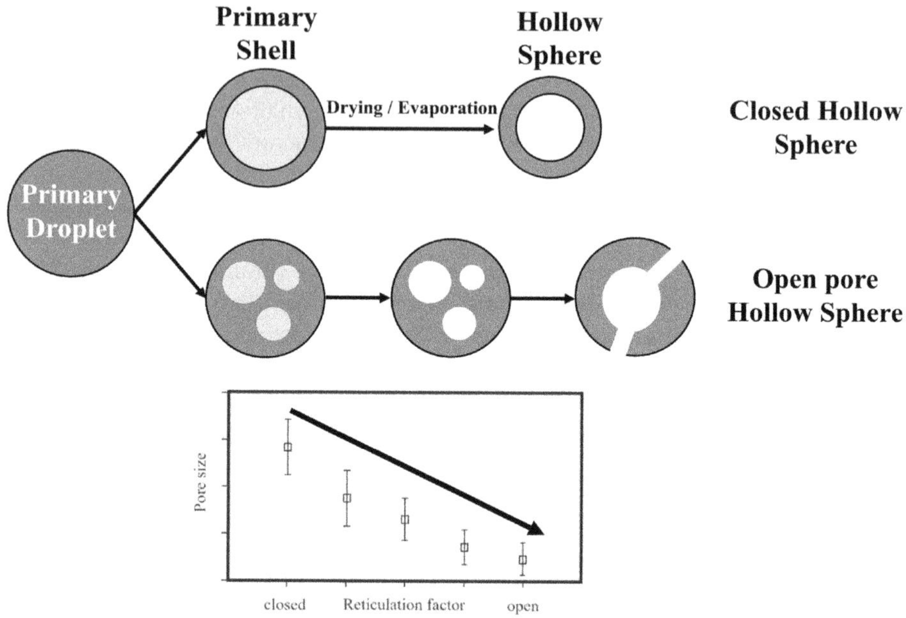

FIGURE 2.8   Open and closed hollow spheres with a possible relation between pore size and permeability. Created, designed, and introduced by the author.

Carbon foams are unique elemental structures with different allotropes. They have the ability to form structures with varying properties through processing and the degree of heat treatment. In this structure, bulk foam is made from carbon or graphite, which are chemically enhanced or heat-treated to create nanostructures, sheet-like structures, tubes, balls, onions, and similar forms. Typical processes include blowing, carbonizing, and thermally treating the foam structures. Carbon foams are porous carbon products containing regularly shaped, predominantly concave, and homogeneously dispersed cells. These foams tend to interact to form a 3D array throughout a continuum material of carbon, predominantly in the non-graphitic state. The final foams are either open or closed-cell products. Specifically, graphite foams are porous graphite products with regularly shaped, concave, and homogeneously dispersed cells. Due to the wide range of structures developed and specific precursors used to fabricate carbon and graphite foams, there is a wide range of physical properties. *Glass foams* have a unique combination of properties, including lightweight, rigidity, compression resistance, thermal insulation, freeze tolerance, nonflammability, chemical inertness, nontoxicity, and water and steam resistance. Glass foams are obtained using gas-generating agents (gasifiers or foaming agents, mostly carbon or carbonaceous substances). Their ground mixtures are heated to a temperature at which gas evolution from the foaming agent occurs within the pyroplastic softened glass particles undergoing viscous flow sintering. The released gas creates a multitude of initially spherical small bubbles under increasing gas pressure, which expand to form foam structures with polyhedral cells. The properties of foam glasses are controlled by the type and quantity of foaming agents, the initial size of glass particles, and the heat treatment regime. The microstructural homogeneity of glass foams is limited to the glass and foaming agent powders. The three main types of glass foams include loose glass foam aggregate, powdered waste glass, and glass foam blocks and shapes.

Three-dimensional periodic structures are composed of interconnected cylindrical rods with applications in advanced ceramic sensors, composites, tissue engineering scaffolds, and photonic materials. The most applicable process is based on colloidal inks developed for the direct assembly of 3D periodic structures. Through well-controlled viscoelastic response, the inks flow through a deposition nozzle and set to facilitate shape retention. The mixture contains high volume fractions of colloidal ceramics to minimize drying-induced shrinkage after assembly; that is, the particle network must be able to resist compressive stresses arising from capillary tension. The colloidal gels used consist of a network of attracting particles capable of transmitting stress above a critical volume fraction, known as the gel point. While the colloid volume fraction within the gelled ink is held constant, the elastic properties of the ink can be controlled by altering the strength of the interparticle attractions. Table 2.4 lists some typical properties of a ceramic/glass foam.

In *connected fibers*, porous fibrous ceramics are used to create fiber felts and mats for applications such as gas transport, gas adsorption, catalyst supports, hot gas filters, composite reinforcement, biomaterials, and acoustic and thermal insulation. The fibers are usually subjected to high-temperature heat treatment stages to fully crystallize the fiber and remove residual porosity. This heat treatment is accompanied by phase changes involving transition oxides. Fibers include alumina, zirconia, mullite, and yttrium aluminum garnet, with an average diameter ranging from 5 nm to 0.1 mm. Widely produced glass fibers are mostly created through melt blowing (silica and glass systems), blowing (alumina–silica), and drawing (drawn alumina–borosilicate). Fibers can be formed as continuous monofilaments, fiber mats, bulk fiber, and similar structures. Continuous monofilaments are monofilaments of continuous blown or drawn fibers, which are usually collected in aligned fiber bundles. These continuous fiber bundles can be chopped into short lengths and mixed with discontinuous fibers to form mats, molded felts, or blocks. Fiber mats are formed by gathering staple fibers that are randomly oriented in the length and breadth directions. The structure is built up in the thickness direction as layers of such deposits are formed atop one another. This arrangement results in anisotropic physical properties

TABLE 2.4

General Properties of Ceramic/Glass Foam

Density (g/cm^3)	<0.5
Porosity (%)	>85
Flexural strength (MPa)	0.5–1
Flexural modulus of elasticity (GPa)	0.6–1.6
Thermal conductivity (W/m K)	~0.05
Sound conductivity loss (dB/100 mm)	25–27

differing in the plane and through-thickness directions. Mats can be used in their as-deposited form or processed into rigid boards by using organic or inorganic binders. They can also be further processed into bulk or milled fiber. Bulk fiber mats are shaped into bulk form via shredding or chopping machines, which produce 3D fiber agglomerates. This chopping process yields a range of fiber lengths.

The *hollow sphere ceramic* structures have applications in refractory thermal insulation, lightweight composites, fiber optic sensors, laser fusion targets, encapsulation, and gas and chemical storage. Ceramic hollow spheres can be processed by spray techniques such as spray drying, spray pyrolysis, coaxial nozzle, the sacrificial core method, the sol–gel/emulsion method, and layer-by-layer deposition on colloidal templates. Spray techniques are most suitable for bulk production and larger diameter spheres. Solution- or colloid-based techniques create hollow spheres with specific morphology and size. Techniques use highly concentrated solutions/colloids where the solvent holds the ceramic particles (slurry or colloid) or dissolves the inorganic salts or chemical compounds to be coated onto the core particles. Through drying the droplets, dried hollow spheres are formed and can be sintered at high temperatures (>500°C). The main characteristics of hollow spheres include sphere size and aspect ratio (the ratio of sphere diameter to wall thickness). The sacrificial core process involves coating a spherical polymer particle (core) with a slurry or solution of ceramic material. After drying the core–shell composite and removing it by chemical dissolution or by heating to a polymer decomposition temperature, the hollow spheres can be calcined and sintered. Polymer spheres can be negatively or positively charged, which enhances adsorption between the sphere and the coating material (ceramic particles in colloidal form or ceramic-producing salt in solution). The hollow spheres are sintered at high temperatures to ensure the attachment of particles in the wall and to create adequate mechanical strength and rigidity in the hollow spheres. Other techniques include mechano-fusion, modified sacrificial core techniques, multiple-emulsion systems (oil/water/oil), ultrasonic spray pyrolysis, emulsion combustion, and micellized diblock copolymer as templates. Spray and coaxial nozzle techniques prepare hollow spheres using various droplet generation techniques with solutions of target material salts or slurries as the starting materials. The solution or slurry is sprayed through a hot chamber or through a coaxial nozzle to generate spherical droplets. In a spray pyrolysis process, hollow spheres can be generated by atomizing an inorganic precursor in solution form, evaporating the solvent during flight, and precipitating inorganic salt on the surface. The remaining solvent should be removed by drying, and the precipitated salt should be pyrolyzed at higher temperatures (>250°C) to form the desired ceramic phase in the shell of a hollow sphere. The pyrolyzed spheres or particles should be sintered at higher temperatures to achieve sufficient mechanical strength. Aerosol techniques, besides fine oxide particle production, can be used to produce encapsulated oil droplets with metal oxide. In spray drying, water or organic-based slurries of ceramic particles are sprayed in the form of droplets into a chamber containing hot air or other inert gases. Spray drying (atomizing and drying) forms hollow particles. Flocculated slurries lead to solid particles, whereas a dispersed suspension leads to hollow particles after spray drying. The evaporation rate, concentration of the solution, and droplet size are the main controlling factors of spray drying. Formation mechanisms include

(i) formation of a surface layer and extraction of the droplet as the evaporating liquid inside the droplet expands; (ii) diffusion of solids back into the droplet slower than the evaporation of the liquid; (iii) flow of liquid with solid along the surface by capillary action; (iv) evaporation of liquids with the purge out of residual air, creating porous hollow spheres.

Many ceramics have been made into hollow spheres including yttria, silica, hematite, rutile, stannic oxide, ferrous-ferric oxide, $Pb(Zr_{0.52}Ti_{0.48})O_3$, cuprous oxide, zircon, ceria, hydroxyapatite, and non-oxides like BN. Spheres are produced with very small diameters (60–120 nm) and wall thicknesses (1–2 nm). The diameter-to-wall-thickness ratio of hollow spheres is the most important parameter that controls their properties. In some applications (i.e., inertial confinement fusion targets), surface roughness is an important factor.

2.5 BULK CERAMICS

The processing of bulk ceramics requires close consideration of raw materials, powder preparation, forming, heat and thermal treatments, and final finishing.

The selection of *raw materials* is indicative of the features of the final product. The typical concerns at this stage include physical and intrinsic properties, alongside the shape and morphology of starting precursors. A detailed list of parameters with preferred conditions is given in Table 2.5. For instance, the purity of the starting ceramic indicates the level of impurities that can segregate at grain boundaries, as well as the likelihood of creep phenomena, sliding of amorphous phases, and a decrease in mechanical properties at high temperatures. The size distribution of starting raw materials specifies the final density (porosity content) and its relation to theoretical density.

Formulation, by adding organic or inorganic additives and dopants to precursor materials, guarantees the powder's behavior during the forming/shaping and final sintering stages. Besides the main ceramics, formulation includes additives like lubricants, binders, plasticizers, flocculants, dispersants, and sintering aids to enhance and stabilize the sintering process.

TABLE 2.5

Optimal Conditions of a Typical Ceramic Bulk

Parameter	Optimal condition
Microstructure	Stable phase
Particle size (d_{50})	fine (micronized)
Particle size distributions	Normal and Narrow distribution
Particle morphology	Regular and spherical
Agglomerates	Loose
Specific surface area (S)	High
Chemical reactivity	High
Chemistry composition	Homogeneous
Impurity content	As low as possible

During the *shaping/forming* stage, powders are molded and treated to achieve specific shapes in a green compact. This process creates shapes and dimensions that are closest to those required for the final product while significantly decreasing open/close spaces and porosities in the ceramic bulk. It enhances mechanical properties by reaching 50%–70% of the theoretical density. The objective of this process, beyond creating a shape for the component, is to obtain a high degree of density in the green compact without cracking and with a homogeneous distribution of porosity. Various techniques can be used to obtain these shapes, including pressing, slip casting, tape casting, injection molding, extrusion, and electrophoresis.

During *pressing*, whether uniaxial, cold isostatic, or hot isostatic pressing, compacting occurs in a die through the application of pressure. In uniaxial pressing, the pressure is unidirectional in a stiff die (hardened steel) under dry or wet conditions (less than or more than 4% humidity), in the presence of a binder to improve particle adhesion and a lubricant to enhance particle sliding. By subjecting the die to ultrasonic pulses during the filling, porosity and defects can be removed, resulting in increased homogeneity. Cold isostatic pressing involves the application of pressure in a uniform manner in three dimensions. A pre-compacted powder is placed in a thermoplastic die and immersed in a hydraulic fluid (glycerine, hydraulic oil, or water) within a pressure chamber. The fluid transfers the pressure uniformly to every part of the die. The die deformation is followed by the compacting of the powder. Additives, such as lubricants and binders, are used for pre-pressing parts. A high pressure of 500 MPa can be applied through cold isostatic pressing for complex shapes. Adding heating to isostatic pressing (hot isostatic pressing) results in high pressure and high temperature to compact the ceramic materials while subjecting them to simultaneous sintering. Through *slip casting*, a stable ceramic suspension is added to a porous chalk die cavity to precipitate and sediment as the suspended particles lose the liquid phase due to die absorption, resulting in a final solid component. To improve particle sedimentation, the process is carried out under pressure, in a vacuum, through centrifugation of the die, or with ultrasounds. *Tape casting* involves a fluid mixture composed of ceramic powder and organic or aqueous additives (solvent, binder, plasticizer, and the like), which is poured onto a moving tape. Due to the presence of two doctor blades, a film of controlled thickness in the 200–1,300 μm range can be produced. Through this method, a variety of supports or multilayered ceramics are manufactured. High-pressure forming of plastic ceramic pastes serves as the basis for *injection molding* and extrusion, with differences in mold placement and type/shape. In injection molding, a ceramic powder mixed with organic binders is heated to a temperature above the softening temperature of the binders and injected into the cavities of the die, where it cools and solidifies. A debonding treatment is necessary to remove excess binders. During *extrusion*, a similar paste is rendered homogeneously into a cylinder in a vacuum, which removes excess gas from the mixture and obtains the desired shape. By exploiting the properties of very fine particles in solution that are electrically charged through ion adsorption on the surface in the electrophoresis process, the application of an electric field causes particle migration to the electrode die of opposite charge, leading to an increase in component thickness that can be further sintered.

The subsequent high-temperature heating treatments (*sintering*) of ceramic powders are carried out to consolidate them into a dense compact part (coordinate variation in the shape of all grains of a compact powder activated by matter transport mechanisms to fill interparticle voids). The manufactured samples possess the desired properties and shapes through compacting and the elimination of pores from the green parts. Sintering conditions require control of process variables such as temperature, holding time, heating/cooling rates, particle dimensions and shapes, and the sintering environment, all of which contribute to the final microstructure. Most properties of ceramics, i.e., mechanical and electrical properties, are influenced by the sintering treatment. To activate mass transfer mechanisms (diffusion or creep), an energy source that sustains the transport (heat, thermal gradient) is required. Depending on composition and the amount of the secondary phases formed during densification, different sintering processes include solid-state sintering, liquid-phase sintering, and vitrification. During *solid-state sintering*, the green part is thermally treated at a temperature range of 0.7-0.9 of the melting points. Solid-state atomic diffusion produces necks between particles and progressively eliminates porosity. *Liquid-phase sintering* includes particulate solids and a coexisting liquid phase during the sintering process, which has resulted from the melting of one component or the formation of a eutectic due to the interaction of two or more components. This liquid can be transient or persistent during sintering, depending on solubility. The process requires a liquid phase at the sintering temperature, wettability of the solid, and solubility of the solid in the liquid. Consequently, the process is designed through three stages: particle rearrangement (the influence of capillary forces and the filling of pores by the liquid phase), solution-reprecipitation (microstructural development, shape accommodation, and grain growth), and solid-state or skeleton sintering (microstructural coarsening by grain boundary diffusion). Vitrification/glazing occurs via filling porosity with a liquid mixture of about 25 vol. % that vitrifies during cooling. The densification process, which progresses through two main mechanisms, is shown in Figure 2.9.

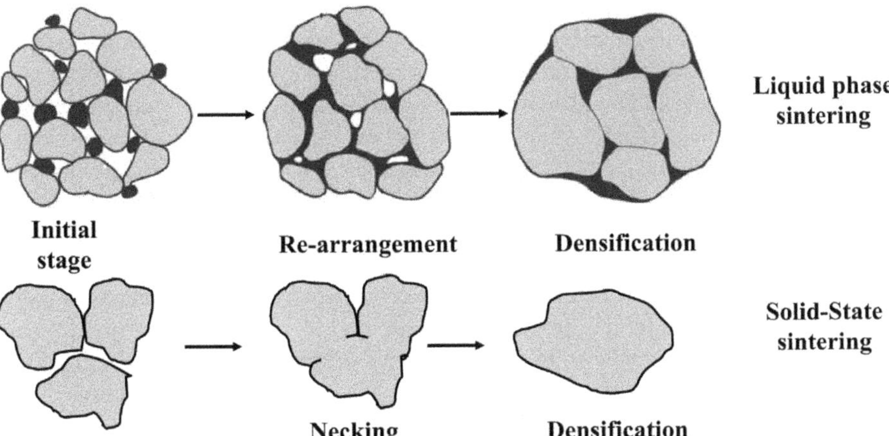

FIGURE 2.9 Liquid-phase sintering and solid-state sintering mechanisms. Created, designed, and introduced by the author.

3 Gas Adsorption
Thermodynamics and Kinetics

Scope

The thermodynamics and kinetics of adsorption are complicated concepts that require close studying and modeling of gas adsorption using basic physical and chemical concepts. The importance of porous substrates as adsorbents of molecular gases is directly associated with their characterization according to thermodynamics and kinetics of adsorption.

3.1 THERMODYNAMICS OF ADSORPTION

The definition of adsorption is the general phenomenon of contact between a fluidized material and the surface of another species, equilibrating along the surface due to attractive interaction forces. Gas-phase adsorption is a key factor in many processes, including multiple gas separation, gas storage, catalysis-assisted chemical reactions, and gas sensing. Adsorption refers to the process of gas molecules attaching to the surface of a solid when in contact with the solid, and this is considered an exothermic process. The term absorption is used when gas molecules in contact with the solid enter the interior of the solid by diffusion. Sorption is a process where both adsorption and absorption occur simultaneously. Desorption refers to the transfer of molecules sticking to the surface of the solid back to its bulk, which is considered an endothermic process. Adsorption phenomena are categorized into physical adsorption and chemical adsorption, also referred to as physisorption and chemisorption. Chemisorption involves electron transfer, leading to bond formation between the sorbate (adsorption surface) and gas molecules. The surface may involve dissociation of the adsorbed species, activation, and can be highly specific to certain adsorbates. Chemisorption requires active sites (binding sites) for chemical reactions. Physical adsorption occurs through van der Waals and electrostatic interactions between the adsorbate and the adsorbent. The van der Waals and electrostatic interaction forces in adsorbed gas-phase systems control the extent of adsorption. A detailed comparison between other specifications of physisorption and chemisorption is provided in Table 3.1 [1].

The main variable in adsorption is the distance (d) of the adsorbate molecule from the adsorbent substrate surface, and the energy of the system is a function of this variable in a simple association of

$$E = E(d) \tag{3.1}$$

DOI: 10.1201/9781003488828-3

TABLE 3.1

Chemisorption versus Physisorption as Adsorption Mechanisms

Chemisorption	Physisorption
Irreversible	Reversible
Specific	Non-specific
High gas temperatures	Low and moderate gas temperatures
Increases with an increase in temperature	Decreases with an increase in temperature
High activation energy	Low activation energy
Formation of a uni-molecular layer	Formation of a multi-molecular layer
Formation of chemical bonds through the interaction between adsorbates and adsorbents	Formation of a adsorption layer without breaking gas molecules

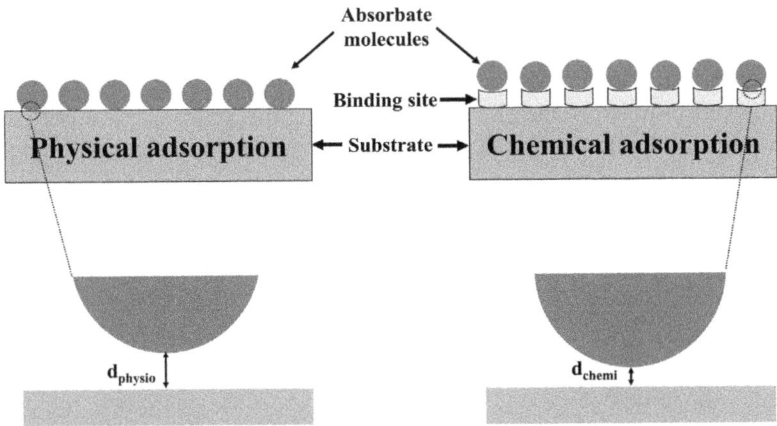

FIGURE 3.1 Graphical presentation of physisorption and chemisorption interactions. Created, designed, and introduced by the author.

This approach assumes a single molecule approaching a clean surface with an understanding of the angular orientation of the molecule, changes in the internal bond angles, bond lengths of the molecule, and the position of the molecule parallel to the surface plane. The final interaction of molecules with the substrate surface depends on the presence of pre-existing adsorbed species, that is, surface impurities and pre-adsorbed molecules (the effect of surface coverage on the adsorption characteristics). The first approach is based on the interaction of an isolated molecule with a clean surface using a simple 1D model. The potential energy (PE) relative to the distance from the adsorbent surface is shown in Figure 3.1. The energy concept is a combination of internal energy, system enthalpy, and free energy of the system. The distinction between physisorption and chemisorption interactions with the surface can be differentiated through the strength of the interaction between the adsorbate particle and adsorbent surface. Generally, the physisorbed particle is farther away from the surface than an identical chemisorbed particle.

In a complete physisorption system, that is, Ar on metal surfaces, the attraction between the adsorbate and surface arises from van der Waals forces. These forces create a shallow minimum in the PE curves at a large distance from the surface ($d > 0.3\,nm$) before an increase in total energy caused by strong repulsive forces that originate from electron density overlap. While weak physical adsorption forces and the associated long-range attraction are formed in the adsorbate/substrate specific surface area (SSA), a chemical bond between the adsorbate and substrate can occur, changing the PE curve by introducing a chemisorption minimum at smaller distances. This phenomenon is regarded as physisorption/molecular chemisorption, which relates to PE changes as a particle approaches the surface to either physisorb or chemisorb. For a typical molecule capable of chemisorption, a combination of two curves with a crossing at the point where chemisorption forces dominate describes the energy profile. Figure 3.2 schematically represents the PE changes during gas particle interaction with an adsorbent surface. At this point, the particle approaches close enough to the surface for the forces of chemisorption to dominate (between d_{physio} and d_{chemi}). The combined PE curves show how physisorption on the surface evolves to chemisorption on the surface and help measure the energy of adsorption, ΔE_{ads}. As the energy of adsorption corresponds to the energy change during this process, it is reported with a negative sign, while the heat of adsorption (Q) is reported with a positive sign:

$$Q = -\Delta E_{ads} \tag{3.2}$$

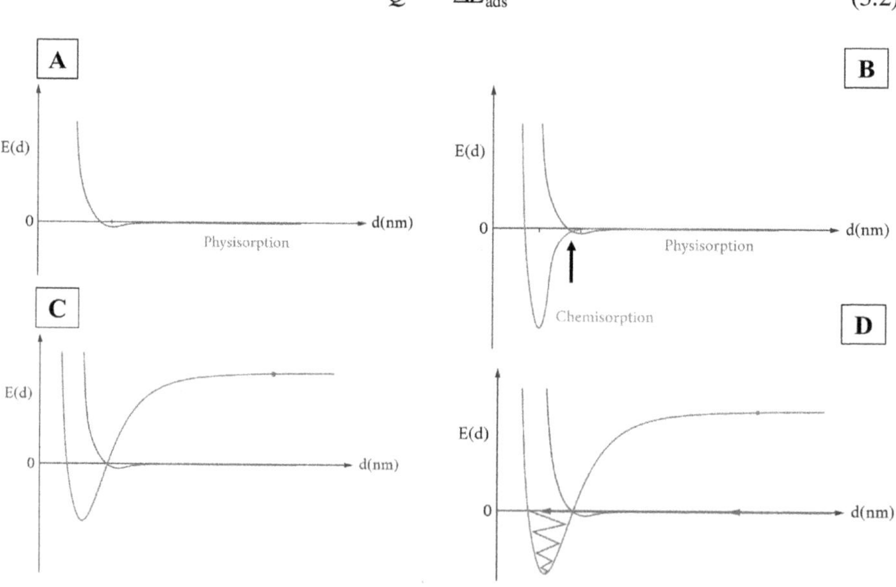

FIGURE 3.2 PE changes for adsorbing gas atom versus distance from surface. A: the PE change of a gas particle approaching an adsorbent surface and physisorbs on the substrate surface; B: the PE change of a gas particle approaching an adsorbent surface and either physisorbs or chemisorbs to the substrate surface; C: the PE change of a gas particle approaching an adsorbent surface and physisorbs to the substrate surface and chemisored to the substrate surface, D: the energy change of a gas particle approaching an adsorbent surface and physisorbed to the substrate surface and chemisored to the substrate surface. Created, designed, and introduced by the author. PE: potential energy.

Atomic adsorption includes simultaneous physisorption and dissociative chemisorption and causes substantial changes in the energy curve of the chemisorption process. Energy interchange involves the dissociation of gas molecules. The molecular gas has two paths for adsorption: (i) energy absorption to separate gas molecules and chemisorption to the surface, and (ii) molecule approach to the surface, physisorption on the surface, and dissociative chemisorption (C). The high energy requirement for molecule dissociation mitigates the molecular breakdown and favors molecule approach to the surface along the physisorption curve. Molecules can proceed directly into the chemisorption well ("direct chemisorption") (D). Alternatively, a gas molecule undergoes transient physisorption, where it can desorb back as a molecule into the gas phase or cross over the barrier into the dissociated chemisorption state.

The PE pathway for physisorption followed by dissociative adsorption without a direct activation barrier includes a free molecule approaching a surface, becoming physisorbed onto the surface, and undergoing dissociative chemisorption (A). In cases where there is a substantial barrier against chemisorption, it can have a major influence on the kinetics of adsorption (B). The activation energies for adsorption and desorption define the energy (or enthalpy) of adsorption (see Figure 3.3, which presents the resulting PE changes for an adsorbent gas particle and its interaction with activation energy).

As the adsorption activation energy (E_{ads}) is smaller than the desorption activation energy (E_{des}), the difference between them (enthalpy of adsorption) is similar to E_{des}. Therefore,

$$E_{des} - E_{ads} = -\Delta E_{ads} \tag{3.3}$$

$$E_{des} \cong -\Delta H_{ads} \tag{3.4}$$

During an adsorption process, one or more components of a gas flow are adsorbed onto the surface of a solid adsorbent. The unsaturated and unbalanced molecular forces present on every solid surface cause adsorption. Additionally, adsorption can be described as a process in which the separation of a substance from one phase is

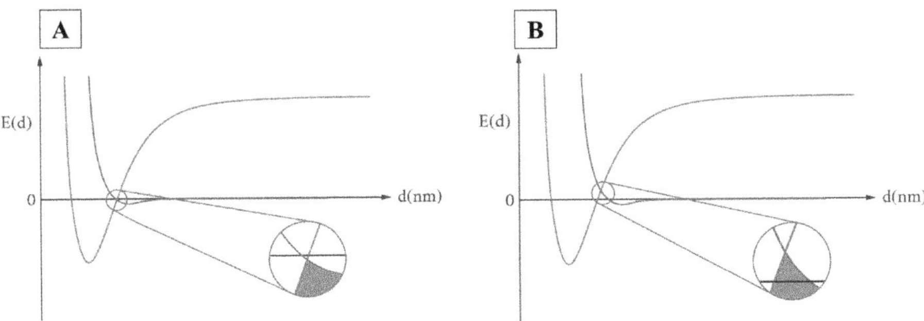

FIGURE 3.3 Effect of activation energy on gas adsorbent–atom interaction with a substrate surface. A: the PE alteration pathway for physisorption, followed by dissociative adsorption without a direct activation barrier; B: the PE alteration pathway for physisorption, followed by dissociative adsorption with a direct activation barrier. Created, designed, and introduced by the author. PE: potential energy.

accompanied by its accumulation on the surface of another. It can also be defined as the accumulation of concentration at a surface. As mentioned, adsorption introduces two types of forces: physical forces and chemical forces. Physical forces in adsorption include dipole moments, polarization forces, dispersive forces, and short-range repulsive interactions. In addition to physisorption by van der Waals forces (dispersion attractive interactions and short-range repulsion), electrostatic contributions, which are significant in adsorbents with ionic structures, should be considered. The energy of adsorption provides valuable information regarding the mechanisms of physisorption. There are two types of contributions: non-specific and specific contributions. The non-specific contributions consist of dispersion and repulsion contribution energies. The specific energy contributions include polarization, field dipole, and field gradient-quadrupole energies.

3.1.1 ISOTHERMS

Isotherms are plots of adsorbent surface coverage by adsorbate as a function of gas pressure at constant temperature. Therefore, an adsorption isotherm is a collection of data indicating the relationship between the pressure of the adsorbate and the quantity adsorbed at a certain temperature per mass or surface area of the adsorbent. Isotherms address the fact that when a gas is in direct contact with a solid surface, an equilibrium will be established between the molecules in the gas phase and the corresponding adsorbate surface species (molecules or atoms). The controlling factors include (i) the relative stabilities of the adsorbed gas-phase units and the surface, (ii) the temperature of the system on both the gas and surface sides, and (iii) the pressure of the gas above the surface. At low pressures, isotherms are measured via the partial pressure as a substitute for fugacity (to fulfill ideal gas thermodynamics conditions for a non-ideal gas from phase equilibrium) since the ideal gas fugacity is equal to the partial pressure. At high pressures, the equilibrium formula would involve fugacity instead of partial pressure as a more accurate substitute term for partial pressure. Isotherms are the most common data presented when describing adsorption capacity. Alternatively, plotting isobars, where the equilibrium capacity is related to temperature at a constant pressure, is common. Similarly, isosteres plots, which provide a relationship between pressure and temperature at a constant adsorbate loading, can also be used. Each of these plots has applications in analyzing porous solid adsorption behavior.

Classification of physisorption isotherms based on the amount of adsorbate relative to the applied pressure is presented in Figure 3.4 (IUPAC-2015).

Type I isotherms of monolayer adsorption are commonly associated with microporous materials with a small relative outer surface area compared to the surface area of internal pores. Type I (a) isotherms, characterized by a sharp increase in uptake at low pressures, result from adsorbents with micropores that have similar dimensions to the adsorbed gas. Type I (b) isotherms are related to micropores larger than the adsorbate gas molecules. When multi-layer adsorption is significant, Type II and Type III isotherms occur. Type II results from strong adsorption, in which a knee point (B) corresponds to monolayer formation. As the saturation pressure is approached, the isotherm diverges due to an infinitely approachable number of

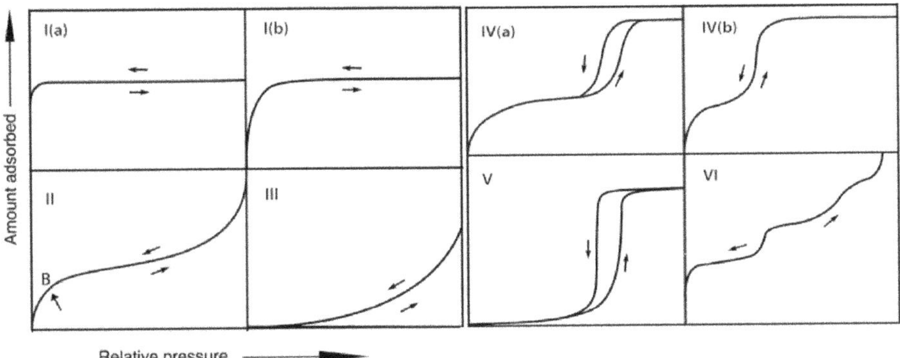

FIGURE 3.4 Different types of physisorption isotherms as observed for different adsorbents. Type I: microporous; Type II: non-porous or macroporous; Type III: non-porous or macroporous with weak interaction; Type IV: mesoporous; Type V: mesoporous with weak interaction; Type VI: layer-by-layer adsorption. Created, designed, and introduced by the author.

stacking layers. Type III represents weak adsorption to the surface, with no discernible inflection point. Type IV isotherms result from mesoporous adsorbents and are divided into Types IV(a) and IV(b). Both Type IV isotherm types reflect monolayer formation followed by a second rise in adsorption quantity. Because the surface of the adsorbents is formed by pore surface area rather than external surface area, they approach a plateau when the pores are filled rather than increasing asymptotically. The difference between Type IV(a) and Type IV(b) isotherms lies in desorption and the hysteresis observed in Type IV(a) adsorption. Type V isotherms are described for adsorbents exhibiting weak adsorbate–adsorbent interactions. They require high pressures to initiate adsorption within the pores, and after initial adsorption, adsorbate–adsorbate interactions become significant. Type VI adsorption isotherms are characterized by stepwise multi-layer adsorption on uniform non-porous surfaces.

3.1.2 Isotherm Models

Adsorption models are developed to extract heats of adsorption from isotherm data.

There are multiple states at the surface and many of these states can coexist. In a system without stacking (no multi-layer adsorption), all surface sites act independently, all adsorption sites are similar, one molecule interacts with one surface site, and adsorbate molecular interactions do not exist [2]. A simple system can be defined in which the quantity of i in the bulk phase ($Q_{i,\text{bulk}}$), quantity of i adsorbed to the surface ($Q_{i,\text{surface}}$), quantity of open surface sites (S_{open}), and the total number of surface sites (S_{total}) can be defined as

$$S_{\text{total}} = S_{\text{open}} + Q_{i,\text{surface}} \tag{3.5}$$

An equation for the rate of adsorption can be derived as

$$\text{Rate}_{\text{adsorption}} = kS_{\text{open}} \cdot Q_{i,\text{bulk}} \tag{3.6}$$

and an equation for the desorption rate is

$$\text{Rate}_{\text{desorption}} = k' Q_{i,\text{surface}} \tag{3.7}$$

where k is the adsorption rate constant and k' is the desorption rate constant. At thermodynamic equilibrium, the rate of adsorption is equal to the rate of desorption; therefore,

$$\frac{Q_{i,\text{surface}}}{S_{\text{total}}} = \frac{K\, Q_{i,\text{bulk}}}{1 + K\, Q_{i,\text{bulk}}} \tag{3.8}$$

where $K = k/k'$. As i species are gases, the concentration can be defined as partial pressure (P_i) and $\dfrac{Q_{i,\text{surface}}}{S_{\text{total}}}$ is the fractional coverage of i on the surface, θ_i (a value between 0 and 1):

$$\theta_i = \frac{KP_i}{1 + KP_i} \tag{3.9}$$

When considering adsorption isotherms, the definition of surface coverage (θ_i), which defines the maximum (saturation) surface coverage of a particular adsorbate on a given surface, should be unity, that is, $\theta_{\text{max}} = 1$. Schematic representation of coverage and pressure dependency is shown in Figure 3.5. Notice the non-linear rectangular hyperbola relation.

The Langmuir isotherm can be derived by treating the adsorption process in a manner similar to other equilibrium processes. The vacant surface site ($S - *$) reacts with gas-phase molecules (M_g) according to the equilibrium reaction:

$$S - * + M_g \rightleftharpoons S - M \text{ and } K = \frac{[S - M]}{[S - *][M_g]} \tag{3.10}$$

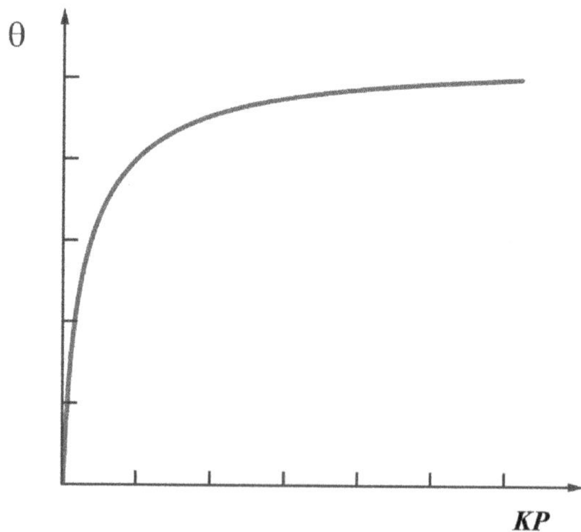

FIGURE 3.5 Non-linear relationship between the fraction of surface coverage and gas pressure during gas atom adsorption. Created, designed, and introduced by the author.

This makes the θ_i formula above as the Langmuir equation. K is an independent constant (independent of θ) if the enthalpy of adsorption is independent of coverage (Langmuir isotherm definition). Plotting θ of versus K_P shows that θ reaches one as pressure increases. The relationship describes the inverse of the fraction of occupied surface sites to the inverse of the gas pressure. This equation describes a Type I(a) isotherm, θ_i term can be expressed in many different forms, such as the ratio of the molar quantity adsorbed over saturated molar capacity ($C_\mu/C_{\mu,\mathrm{max}}$), the volume adsorbed over the maximum volume that may be adsorbed, or the adsorbed quantity over the maximum adsorbed quantity that may be adsorbed ($Q_i/Q_{i,\mathrm{max}}$).

Hysteresis in adsorption behavior occurs when adsorption proceeds by a different process than desorption, resulting in adsorption and desorption equilibrium loadings that deviate from each other. The amount of adsorption on a solid at a certain pressure depends on the isotherm loading point, which was reached by either the adsorption or desorption step. Different types of hysteresis are shown in Figure 3.6. The Type H1 loop is related to materials that exhibit a narrow range of uniform mesopores, while Type H2 loops are shown in more complex pore structures. The steep desorption branch in H2(a) loops can be attributed to pore blocking/percolation in a narrow range of pore necks or to cavitation-induced evaporation. The Type H2(b) loop is associated with pore blocking, where the size distribution of neck widths is larger. The Type H3 loop has two features: (i) the adsorption branch resembles a Type II isotherm and (ii) the lower limit of the desorption branch is located at the cavitation-induced P/P^0. H3 loops are given by non-rigid aggregates of plate-like particles (e.g., certain clays), where the pore network consists of macropores that are not completely filled

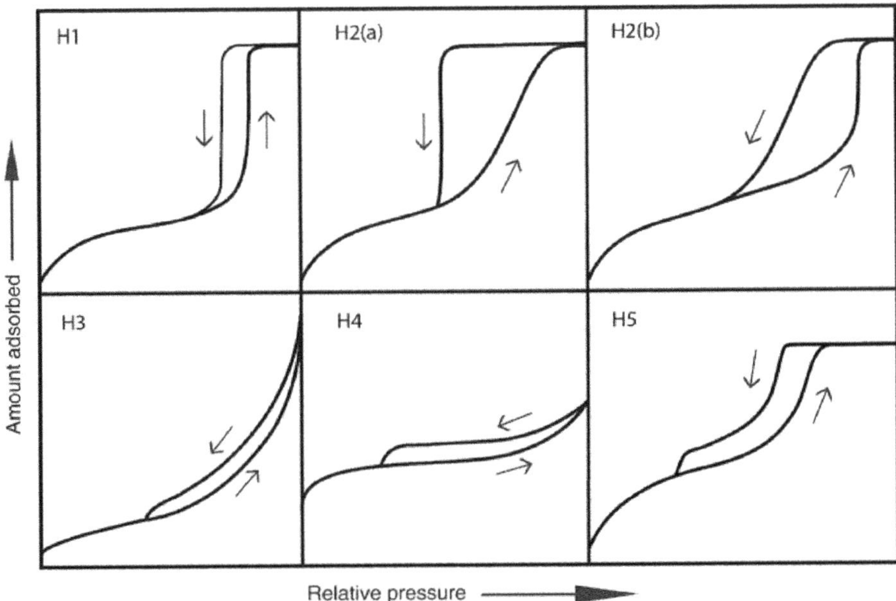

FIGURE 3.6 Types of adsorption isotherm hysteresis loops. Created, designed, and introduced by the author.

with pore condensate. The H4 loop has an adsorption branch that is a composite of Types I and II, with more uptake at low P/P^0 being related to the filling of micropores. The H4 loops are associated with aggregated crystals, mesoporous, and micro-mesoporous carbons. The Type H5 loop has a distinctive form associated with certain pore structures containing both open and partially blocked mesopores.

For monolayer adsorption in a circular pore,

$$\frac{P}{P^0} = \exp\left(-\frac{2\gamma V_m}{RT} \cdot \frac{1}{r_m}\right) \tag{3.11}$$

where P^0 is the saturated vapor pressure, P is the pressure, V_m is the molar volume of the liquid, γ is the surface tension, r_m is the mean radius of curvature of the interface, T is the temperature, and R is the ideal gas constant. Negative gas adsorption is identified when, at a certain temperature, the absolute amount of substance in the adsorbed phase decreases with increasing pressure of the adsorbate, resulting in sections of the isotherm with a negative slope. During the examination of adsorption isotherm data, scanning curves are essential. Typical scanning curves are generated by reversing the adsorption/desorption process over the pressure range, where pores are filled or emptied by condensation or evaporation. The concepts of absolute, excess, and net adsorption are additional terms for defining an adsorbent. In a system of rigid microporous solid with a fixed volume (V_s) of the porous solid and the micropore volume, the total number of moles of A (adsorbate) and S (solid) can be estimated as

$$n_{\text{total}} = n_A + n_S \tag{3.12}$$

In an absolute adsorption system, the solid should be removed, and therefore,

$$n^{\text{abs}} = n_{\text{total}} - n_S = n_A \tag{3.13}$$

In net adsorption, the moles in a fluid at the same pressure and temperature of the system with an equilibrium concentration as the adsorbed phase, occupy the volume of the system:

$$n^{\text{net}} = n^{\text{abs}} - V_s C = n_A - V_s C \tag{3.14}$$

For the definition of excess adsorption, the non-accessible volume (V_{NA}) is defined as follows:

$$n^{\text{ex}} = n^{\text{abs}} - \left(V_s - V_{\text{NA}}\right)c = n_A - \left(V_s - V_{\text{NA}}\right)c \tag{3.15}$$

In a microporous solid, the non-accessible volume can be defined as (i) the geometric volume of the solid, (ii) the volume not accessible to the smallest adsorbate (or the adsorbate for a pure component), (iii) the volume not accessible to a certain molecule. These definitions are essential for calculations of Henry's constants [3].

By defining a system composed of a two-dimensional interface (the adsorbent sites) and a gaseous phase above, the change in Helmholtz free energy of the surface of the adsorbent sites (dF_s) can be defined as follows:

$$dF_s = -S_s dT - P dV_S - \mu_s dn_s + \mu_i dn_i \tag{3.16}$$

where S_S is the entropy of the surface sites, V_S is the volume of the adsorbent sites, μ_i is the chemical potential of the adsorbing species i, n_i is the moles of adsorbing species i, μ_s is the chemical potential of adsorbent sites, and n_s is the moles of adsorbent sites. By assuming constant temperature (isotherms), dV_S can be estimated as negligible, while n_s is directly proportional to the surface area A and may be expressed as the spreading pressure of species i (π_i). The spreading pressure can be defined as the reduction in surface tension on the surface upon the addition of the adsorbate. Therefore, dF_s can be defined as follows:

$$dF_s = -\pi_i dA + \mu_i dn_i \tag{3.17}$$

It can be shown that

$$Ad\pi_i = n_i d\mu_i \tag{3.18}$$

As μ_i for the adsorbed phase must be equivalent to μ_i in the gaseous phase, by estimating the gaseous phase as an ideal gas, $d\mu_i$ is defined as follows:

$$d\mu_i = \frac{RTdP}{P} \tag{3.19}$$

By combination, the general Gibbs isotherm can be defined as follows:

$$\left(\frac{d\pi_i}{dP}\right)_T = \frac{n_i}{A} \cdot \frac{RT}{P} \tag{3.20}$$

With an equation of state for the condensed phase, the generalized Gibbs isotherm equation can be solved for the adsorbed quantity as a function of pressure. By assuming an equation of state comparable to the ideal gas law, π can be estimated as follows:

$$\pi = \frac{n_i RT}{A} \tag{3.21}$$

And

$$\left(\frac{d\pi_i}{dP}\right)_T = \frac{\pi}{P} \tag{3.22}$$

by integration

$$\pi = C(T)P \tag{3.23}$$

where $C(T)$ is the integration constant, which is a function of temperature. By defining that n_i / A is equivalent to θ_i under the low surface coverage limit, a linear isotherm is defined as follows:

$$\theta = \frac{C(T)P}{RT} \tag{3.24}$$

where $C(T)/RT$ is followed by a single temperature-dependent adsorption constant K (called Henry's constant):

$$\theta = K(T)P_i \tag{3.25}$$

This linear isotherm for the limit of surface coverage expresses the limit of adsorption near infinite dilution following Henry's law.

Another widely used application of isotherm analysis is pore size distribution. The pore size distribution of an adsorbent can be determined from the hysteresis loop (the desorption or adsorption branch of the isotherm) when pore filling and emptying occur in a controlled manner. It should be emphasized that these pore size distributions arise by relating the pressure at which uptake occurs on the isotherm to theoretically determined geometries. The pressure of the plateau in absorption isotherms for many gas compounds (i.e., hydrides) is always higher than the pressure of the plateau for desorption at any temperature, but the degree of hysteresis varies between compounds, with absorption and desorption plateaus both representing metastable states. The plateau pressure hysteresis is not very significant and barely prevents hydrogenation at the charging pressure or dehydrogenation at its delivery pressure. However, minimization of hysteresis is an ideal practice, and its existence is a factor to consider in the characterization of hydrides for storage applications.

Many empirical isotherms have been introduced for more precise estimation of adsorption behavior. Monolayer isotherms, such as the Freundlich isotherm, are basic empirical models using fitting parameters K and n to define the relationship of $Q_{i,\text{adsorb}} = KP^{1/n}$. This empirical model shows that P increases with $Q_{i,\text{adsorb}}$ without bounding and also does not follow Henry's law behavior at low pressures. Despite these limitations, the n factor captures the heterogeneous behavior of adsorbents at intermediate pressures and was combined with the Langmuir equation by Sips to form the Langmuir–Freundlich isotherm (the Sips isotherm):

$$\theta = \frac{Q_i}{Q_{i,\max}} = \frac{KP^{1/n}}{1 + KP^{1/n}} \tag{3.26}$$

While this equation captures inhomogeneous behavior, it shows good conformity at high pressures without fulfilling Henry's law behavior at the low-pressure limits.

A conformity model for an isotherm that considers different types of available sites for binding on the adsorbent is the dual-site Langmuir–Freundlich isotherm:

$$Q_i = \frac{N_1 k_1 p^{n1}}{1 + k_1 p^{n1}} + \frac{N_2 k_2 p^{n2}}{1 + k_2 p^{n2}} \tag{3.27}$$

which is composed of the maximum adsorption capacity at site i (N_i), the adsorption constant for site i (k_i), and the Freundlich deviation from Langmuir adsorption for site i (n_i). This isotherm can capture inhomogeneous behavior and is useful for modeling adsorbents when the low coverage limit is not a concern. A well-established isotherm that can be used to model low-pressure inhomogeneous behavior of Henry's law is the Toth isotherm expressed as follows:

$$C_\mu = C_{\mu s} \frac{bP}{[1 + (bP)^t]^{1/t}} \tag{3.28}$$

Here, b is the coefficient of adsorption affinity at low pressure, t is the system heterogeneity, $C_{\mu s}$ is the adsorption capacity, C_μ is the amount adsorbed, and P is the pressure. At $t = 1$, the equation reduces to the Langmuir equation. The terms b, t, and $C_{\mu S}$ have a temperature dependence, as given below:

$$b(T) = b_0 \exp\left(\frac{Q}{R_g T_0}\left(\frac{T_0}{T} - 1\right)\right) \tag{3.29}$$

$$t(T) = t_0 + \alpha\left(1 - \frac{T_0}{T}\right) \tag{3.30}$$

$$C_{\mu S}(T) = C_{\mu S,0} \exp \chi\left(1 - \frac{T_0}{T}\right) \tag{3.31}$$

where T_0 is the reference temperature, Q is the heat of adsorption at zero coverage, b_0 is the adsorption affinity at T_0, $C_{\mu S,0}$ is the adsorption capacity at T_0, t_0 is the heterogeneity of the system at T_0, and α and χ are fitting parameters.

While monolayer adsorption isotherms conform well to Type I(a) and Type I(b) behaviors, they fail to describe other isotherm types. Models quickly become more complex than the monolayer adsorption models when trying to capture the complexities of adsorbate interactions and adsorbent geometry. The multi-layer isotherms are more complex than the monolayer adsorption models because they capture the complexities of adsorbate interactions and adsorbent geometry, such as adsorption into cage-like structures (i.e., networks composed of small apertures with large pore volumes) and micropore filling. The mono to multiple layer adsorption stacking is shown schematically in Figure 3.7.

The Brunauer–Emmett–Teller (BET) isotherm allows for multi-layer adsorption modeling as follows:

$$v = \frac{v_m c P}{(P_0 - P)\left[1 + \left(\frac{(c-1)P}{P_0}\right)\right]} \tag{3.32}$$

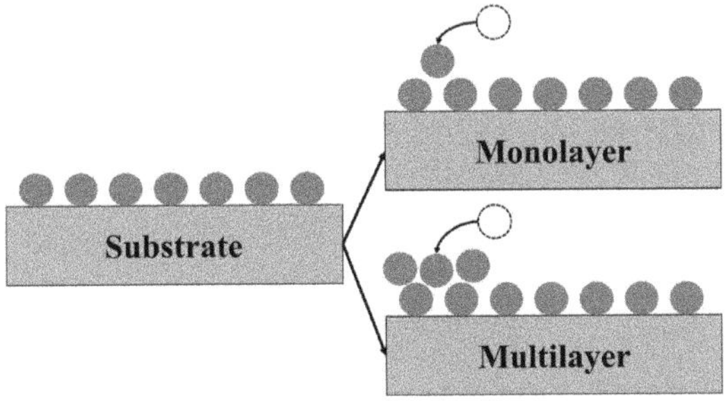

FIGURE 3.7 Monolayer (BET) and multi-layer adsorption. In monolayer adsorption, all the adsorbed molecules are in contact with the surface layer of the adsorbent. In multi-layer adsorption, the adsorption space accommodates more than one layer of molecules and not all adsorbed molecules are in contact with the surface layer of the adsorbent. Created, designed, and introduced by the author.

in which v is the amount adsorbed in volume, v_m is the monolayer amount adsorbed, c is the BET constant, and P is the total pressure, with P_0 as the saturation vapor pressure. The BET approximation for surface area is limited to the assumption that gas molecules stack directly on top of each other; however, they are more likely to pack in a triangular fashion, known as the Halsey approximation, which describes how complete a monolayer should become before a second layer starts forming. The Dubinin–Astakov isotherm is a general micropore filling model defined as follows:

$$v = \frac{v_m c P}{(P_0 - P)\left[1 + (c-1)\dfrac{P}{P_0}\right]} \tag{3.33}$$

As a result,

$$\theta = \frac{v}{v_0} = \exp\left(-\left(\frac{A}{E}\right)^n\right) \tag{3.34}$$

where A is the adsorption potential, E is the characteristic energy for adsorbate interactions with the bulk gas of a similar kind in micropores, and n is the pore size distribution. For an ideal gas, the adsorption potential A is introduced as follows:

$$A = RT \ln\left(\frac{P_0}{P}\right) \tag{3.35}$$

in which P_0 is the vapor pressure, P is the equilibrium partial pressure, and n is an empirical constant. For $n=2$, isotherm changes to its original derivation, which captures the typical pore size in ceramic substrates. At $n=3$, the adsorbent can be considered to have a normal pore size distribution in highly organized adsorbents. Models such as Virial expansion isotherms are applicable means for describing general isotherms through their potential to fit experimental data using factors such as Henry's constant and heat of adsorption. Overall,

$$\ln(P) = \frac{1}{T}\sum_{i=0} a_i q^i + \sum_{j=0} b_j q^i + \ln(q) \tag{3.36}$$

where a_i and b_j are fitting factors that need to be added for the isotherm to fit the experimental data, starting from $i=0$ and $j=0$, respectively. The Henry's law constant can be expressed through the follows equation:

$$K = \exp(-b_0)\exp\left(\frac{-a_0}{T}\right) \tag{3.37}$$

Besides these stacking models of adsorption, the enthalpy of adsorption (ΔH), known as the heat of adsorption or isosteric heat (isosteric meaning constant coverage), is the binding energy of adsorbents during the adsorption process. The importance of the enthalpy of adsorption is related to (i) indicating the energy required during desorption cycles to regenerate the adsorbent, (ii) describing the heat generated during adsorption as the limiting factor in adsorption kinetics when heat dissipation is scarce, (iii) indicating the magnitude of the heat of adsorption as a binding determinant, and (iv) relating changes in ΔH to the surface coverage of heterogeneous surfaces where different parts of the adsorbent have different binding energies, or adsorption occurs through different mechanisms as a function of adsorbed content, such as a transition from multi-layer adsorption to monolayer adsorption. The values of the heat of adsorption are below 40 kJ/mol as an indication of van der Waals' binding force (physisorption). Values greater than 40 kJ/mol indicate chemical bond formation (chemisorption). The enthalpy of adsorption can be derived from Van't Hoff's equation given as follows:

$$\left(\frac{\delta \ln (P)}{\delta \left(\dfrac{1}{T} \right)} \right) = -\frac{\Delta H}{R} \tag{3.38}$$

where P is the pressure, T is the temperature, and R is the gas constant. When the temperature-dependent isotherm fits the experimental data, it is possible to solve the isotherm equation directly for ΔH as a function of surface coverage.

$$\Delta H = -R \sum_{i=0}^{n1} a_i v^i \tag{3.39}$$

In isotherms with basic temperature-dependent terms, ΔH can be interpolated from isotherms at multiple temperatures using a plot of $\ln(P)$ versus T^{-1} as a straight line with a slope of $-\Delta H/R$ (a Van't Hoff plot):

$$\ln \frac{k_2}{k_1} = -\frac{\Delta H^0}{R} \left[\frac{1}{T_2} - \frac{1}{T_1} \right] \tag{3.40}$$

It is recommended that three different temperatures spaced 10°C apart be used for calculations. The concept is shown schematically in Figure 3.8. In a similar approach, the enthalpy of complex formation or decomposition with a gas (i.e., hydride) is regularly calculated from pressure–composition isotherms using a graph of the natural log of the plateau pressure versus inverse temperature. The typical absorption–desorption pressure–composition–temperature (P–C–T) curves and kinetics are measured by a setup based on the principle that a closed and sealed system contains an adsorbent, either in a gaseous state or carried within the solid surface. If the volume is calibrated and kept precisely, the amount of gas can be calculated by measuring the respective temperatures and pressures and using a real gas equation of state.

The gas absorption temperature, which is the lowest temperature (T_L) in the process, and the hydrogen desorption temperature, which is the highest temperature (T_H) in the process, must be discovered. Then, the absorption–desorption P–C–T curves of the hydrogen storage alloy at the respective temperatures of T_L and T_H must be acquired. The typical absorption or desorption P–C–T curve includes a plateau region (AB). The plateau pressure of the absorption side of the curve is greater

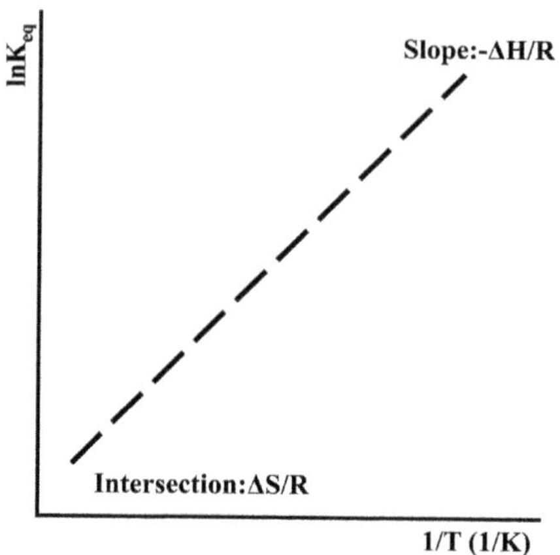

FIGURE 3.8 Graphical presentation of Van't Hoff's equation. Created, designed, and introduced by the author.

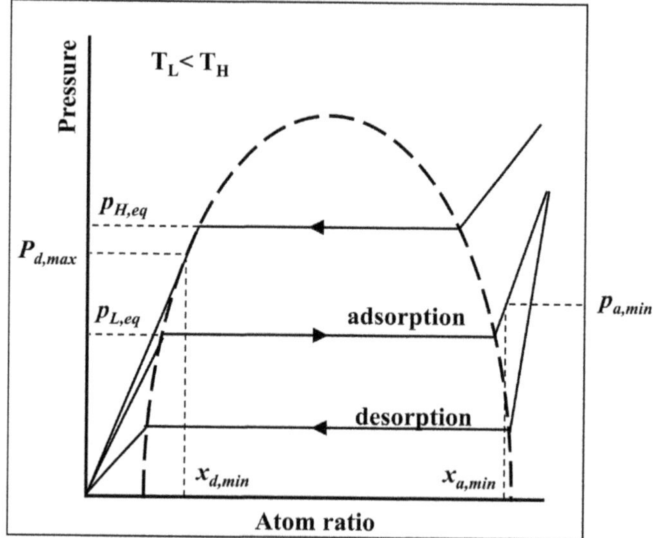

FIGURE 3.9 Effect of temperature on isotherm plateau pressure and phase transitions. Created, designed, and introduced by the author.

than that of the desorption side (see Figure 3.9). The absorption or desorption plateau pressure increases with an elevation in test temperature. Typically, pure thermal cycling adsorption/desorption consists of absorption at low temperature and desorption at high temperature. Hence, it is necessary to control the volume of the system to maintain the pressure of the system after absorption at temperature (T_L) to be greater than $p_{L,eq}$ and the pressure in the system after desorption at T_H temperature to be less than $p_{H,eq}$.

To study the adsorbent–adsorbate interactions, the analysis of the isotherm must be conducted at very low surface coverage. This condition allows for the elimination or minimization of adsorbate–adsorbate interactions. Therefore, at higher substrate coverage, an additional (self-potential) term must be added to the energy equations to compensate for these interactions. As the adsorption energy is governed by both the adsorbent and the adsorbate, the magnitude of the non-specific contribution depends on the polarization ability of the adsorbate and the adsorbent, as well as the density of the force centers in the outer part of the adsorbent (i.e., the surface layer). The polarity of the adsorbate molecule is only important when the specific contribution is significant.

Adsorbate molecules entering small pores with molecular dimensions result in a considerable increase in adsorption energy compared to the adsorption energy associated with the physisorption of the same molecules on the corresponding open surface. This increase is due to the additive nature of the molecular interactions.

The enthalpy of adsorption provides a measure of the strength of the interaction between a molecule and the adsorbent surface or pore structure. Micropores result in the overlap of potential fields from opposing pore walls, leading to a subsequent increase in the density of the adsorbed gas at any temperature and pressure. The enthalpy of adsorption can be defined through the transformed molar surface excess enthalpy, transformed differential surface excess enthalpy, transformed differential enthalpy, transformed integral molar enthalpy, differential surface excess enthalpy, and the differential enthalpy of adsorption. The differential enthalpy of adsorption can be calculated from the measurement of at least two isotherms at different temperatures using the isosteric method. The differential enthalpy of adsorption, determined for a particular uptake (surface coverage), is known as the isosteric enthalpy of adsorption (ΔH_{iso}). In ΔH_{iso} estimation, the pressure at which a certain amount of gas is adsorbed at different temperatures is essential. One method is to apply the following expression (Clausius–Clapeyron equation):

$$-\Delta H_{iso}(n) = -R \cdot \ln\left(\frac{P_2}{P_1}\right)\frac{T_1 \cdot T_2}{T_2 - T_1} \tag{3.41}$$

where T_2, T_1 are two close measurement temperatures, P_1 and P_2 are the pressures at which a given quantity of gas is adsorbed, and R is the universal gas constant. To interpolate between data points, it is necessary to fit the adsorption data to an appropriate adsorption equation. The Langmuir equation used to fit the adsorption data shows some scattering at higher pressures.

The most important and practical application of adsorption data is in the practical storage properties of solid substrates. The practical storage properties include reversible storage capacity, which encompasses the excess and absolute adsorbed quantities for adsorbents, long-term cycling stability, and the resistance of a material to gaseous impurities. The reversible storage capacity of an adsorbent is the quantity of gas sorbed and desorbed between the lower and upper operating pressures. This factor is technologically as important as total or maximum storage capacity. The difference between these two values depends on the uptake behavior of the substrate. An important aspect includes the isotherm shape and limits, as they determine the pressure range over which the majority of the reversible uptake occurs. It can also be affected by the rate of gas uptake and release, as some of the stored gas cannot be released in a practical timeframe due to kinetic limitations. As a result, this also prevents full gas recharging. Regarding isotherm shape, a substrate with a single plateau in its isotherm and microporous substrates that exhibit typical Type I adsorption isotherm behavior have significant reversible storage capacities. In the case of the plateau, the reversible storage capacity can be determined by the width of the plateau in the pressure–composition isotherm. Since this is the main part of the isotherm used to indicate storage, there is a small amount of uptake in either single-phase region, and the reversible gas storage capacity is close to the maximum content on the substrate. The temperature of the sorption process affects the reversible capacity by increasing the absorption plateau pressure with an increase in temperature. The schematic behavior of reversible storage and the effect of increasing temperature is shown in Figure 3.10.

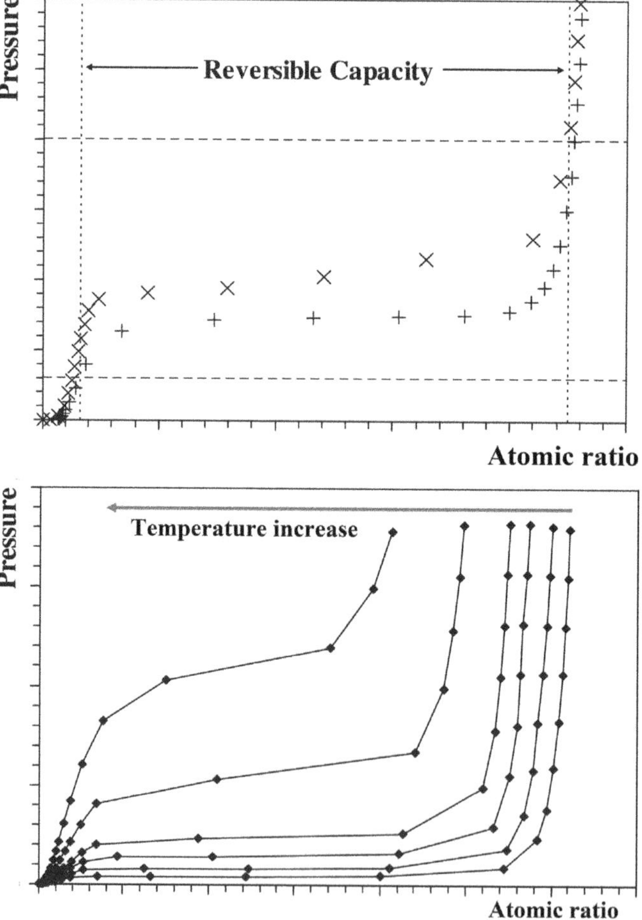

FIGURE 3.10 Reversible adsorption and how it is affected by temperature. Created, designed, and introduced by the author.

In a typical microporous adsorbent, the substrate shows Type I adsorption iso-therm behavior at low temperatures. As defined for Type I behavior, the equilib-rium uptake initially increases significantly with pressure up to a plateau where gas uptake saturates and does not increase further to any significant extent (as shown in Figure 3.11). The reversible capacity from these isotherms at a particular temperature depends mainly on the lower storage pressure and to a lesser extent on the upper pres-sure. The reversible adsorption capacities in these pressure ranges are considerably less than the maximum storage capacity and depend significantly on the operating pressure range. Although the upper and lower operating pressures are essential, the temperature of operation has a significant effect on the reversible storage capacity, as the uptake at a certain pressure generally decreases significantly with increasing temperature.

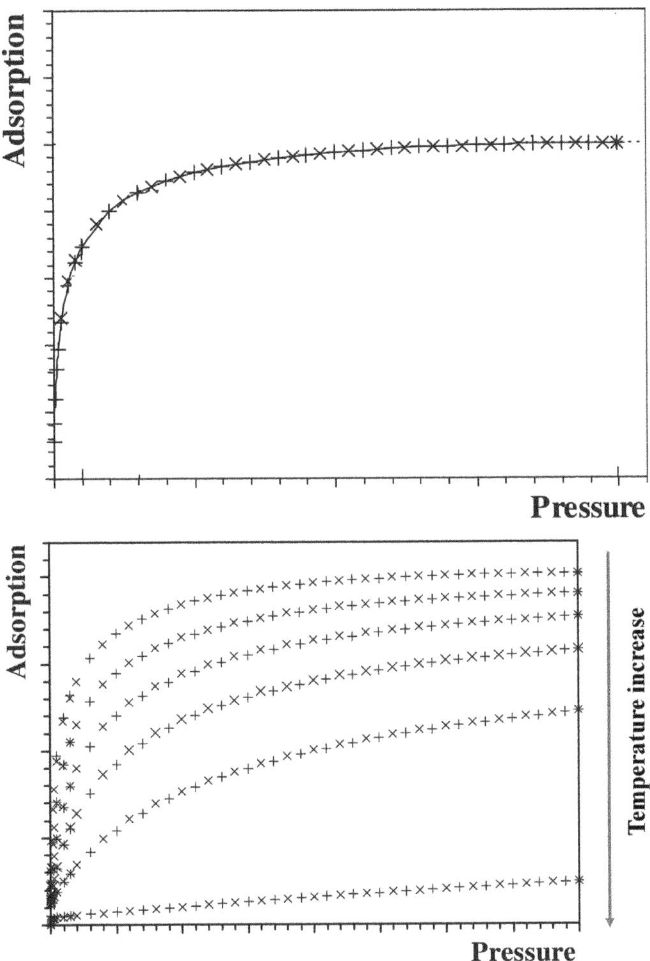

FIGURE 3.11   Gas adsorption uptake plateau and how it is affected by temperature. Created, designed, and introduced by the author.

The gravimetric and volumetric storage capacities are two indicative factors in adsorption definitions. The gravimetric storage capacity is the amount of gas adsorbed per unit mass of substrate. It can be defined as a quantity and is calculated from the ratio of hydrogen mass adsorbed by the adsorbent to the mass of substrate material, including the absorbed hydrogen (capacity in wt.% or $C_{wt.\%}$):

$$C_{wt.\%} = \left(\frac{\left(\dfrac{G}{M}\right)M_g}{M_{substrate} + \left(\dfrac{G}{M}\right)M_g} \right) \tag{3.42}$$

or

$$C_{wt.\%} = \left(\frac{n_a M_g}{n_{substrate} M_{substrate} + n_a M_g} \right) \tag{3.43}$$

where G/M is the host gas-to-materials atomic ratio, M_g is the molar mass of gas, $M_{substrate}$ is the molar mass of the host material, n_a is the excess adsorption in moles, and $n_{substrate}$ is the number of moles of host material, reported in typical units for the measured quantity (the excess adsorption) in the number of moles of adsorbate per unit mass of adsorbent. The volumetric storage capacity is defined as the amount of hydrogen stored per unit volume of the substrate. This quantity can refer to the amount of gas stored per unit volume of the bulk adsorbent or compound. The possible expansion of the substrate lattice due to gas absorption must be neglected, as many adsorbed gas atoms can enter the bulk of the substrate material and be contained within the boundaries of the crystal lattice. It should be mentioned that under real conditions, some lattice expansion is conceivable. The geometric density of the host substrate can be used if the lattice (unit cell) volume does not change significantly with hydrogen absorption. Nevertheless, it should be assumed that the adsorbed phase occupies the entire volume of the pores. Excess adsorption, or *Gibbs excess* or Gibbsian surface excess in gas adsorption, is the difference between the amount of gas phase that would be present in the equivalent volume of the adsorbed phase in the absence of adsorption and the actual total amount in the phase. The total quantity is the total or absolute adsorption. The gravimetric adsorbed hydrogen density $(\rho(x))$ relates to total adsorbed mass (m_t) as follows:

$$m_t = \int_{V0}^{V} \rho(x)dV \tag{3.44}$$

$$\rho_g = \lim_{x \to \infty} \rho(x) \tag{3.45}$$

where V indicates the volume of gas in the adsorption state. The constant bulk density of gas (ρ_g) is achieved at a far distance (x_a) from the adsorbent surface. The excess adsorbed mass (m_e), as the difference between the total adsorption and the mass of gas present in the volume of the adsorbed phase $(V(A) = V_a)$ in the absence of gas–surface interactions, is derived from the difference between $\rho(x)$ to ρ_H as follows:

$$m_e = \int_{V0}^{V} \left(\rho(x) - \rho_g\right) dV \tag{3.46}$$

to make

$$m_e = m_t - \rho_g V_a \tag{3.47}$$

The volume of the adsorbed phase (V_a) is necessary for indicating $|m_e - m_t|$ as most experimental adsorption measurements only indicate the m_e value. As ρ_g becomes more significant with increasing gas pressure, the $|m_e - m_t|$ value increases. If the adsorbate phase density is larger than the bulk gas-phase density, $|m_e - m_t|$ approaches zero. As the adsorbate can condense to a liquid-like state on the surface at subcritical adsorptive temperatures and V_a is more clearly defined, the $m_e \approx m_t$ condition is dominant. Under these conditions, the system total volume (V_t) is a combination of solid adsorbent volume (V_s) including any inaccessible pore volume (V_a), the adsorbed-phase volume, and the bulk fluid (gas)-phase volume (V_f) as follows:

$$V_t = V_s + V_a + V_f \tag{3.48}$$

and the reference mass quantity is

$$m_r = \rho_H \left(V_t - V_s\right) \tag{3.49}$$

The dead volume of the adsorption system $\left(|V_t - V_s|\right)$ is filled with the mass of the gas phase at a density of ρ_g without adsorption. This volume indicates the Gibbs dividing surface $(x=0)$ (see Figure 3.12). By estimating the total mass of gas present in the adsorption measurement system (m_s), excess adsorption mass can be derived as follows:

$$m_e = m_s - m_r \tag{3.50}$$

In this approach, gas adsorbates in the non-adsorbed phase and in the adsorbate phase are considered similarly. Therefore,

$$m_t = m_e + \rho_g V_a \tag{3.51}$$

$$V_a \approx V_p$$

for estimated V_a and $\rho_g V_a$ with V_p theoretical pore volume. This approach assumes that adsorbed-phase volume is independent of pressure and relates to the pore volume of the substrate (total pore volume). The main estimations can be categorized into (i) assuming a constant adsorbed-phase volume and (ii) assuming a constant phase density. The total pore volume (V_p) is based on the first assumption. By using the molecular diameter of the adsorbate and the SSA of the adsorbent, the effective thickness can be estimated for adsorbents with clear SSA, not including microporous surfaces.

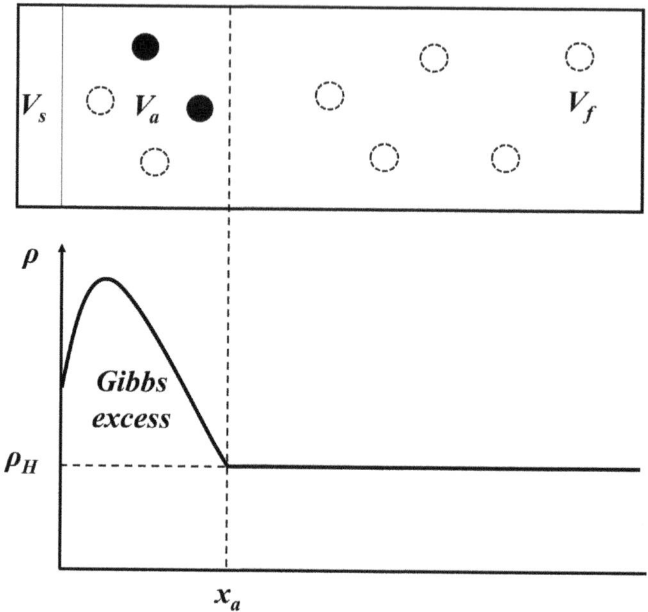

FIGURE 3.12 Graphical representation of Gibbs excess and total adsorbed quantities. The position of $x=0$ the Gibbs dividing surface. Black circles are molecules related to the Gibbs excess, and white circles are molecules related to the remaining curve area. Created, designed, and introduced by the author.

By estimating a constant value for the mean adsorbate phase density (ρ_a) at a certain temperature, the V_a value can be calculated as follows:

$$V_a = \frac{m_t}{\rho_a} \tag{3.52}$$

Other approximations such as critical density approximation (ρ_L^b) and liquid density approximation (ρ_c), use substitutions for ρ_a derived as follows:

$$\rho_a = \frac{\rho_L^b}{\exp\left(\alpha\left(T - T_b\right)\right)} \tag{3.53}$$

$$\rho_a = \frac{M}{b} = \frac{MRT_c}{8P_c} \tag{3.54}$$

where ρ_L^b is the density at T_b boiling point, α is the thermal expansion coefficient for superheated liquid, b is the van der Waals constant, M is the molar mass, R is the universal gas constant, T_c is the critical temperature, and P_c is the critical pressure. ρ_g can be related to fugacity (f) as follows:

$$\rho_g = \frac{Mf}{RT} \tag{3.55}$$

where other approximations can be rewritten as follows:

$$m_t(f) = m_e(f) + \left(\frac{Mf}{RT}\right)\left(\frac{m_t}{\rho_a}\right) \tag{3.56}$$

$$m_t(f) = \frac{m_e(f)}{\left(1 - \dfrac{Mf}{RT\rho_a}\right)} \tag{3.57}$$

In difficult cases, estimating the exact volume of the system can be challenging. The gravimetric data can be derived by combinations of experimentally measured mass (m_{exper}), the mass of the adsorbent or solid (m_s), mass of the adsorbed phase ($\rho_g V_d$), and a correction as follows:

$$m_{\text{exper}} = m_s + m_t - \rho_g V_d \tag{3.58}$$

The involved volume is a combination of the volume of the solid (V_s), including any inaccessible pores, and the volume of the adsorbed phase (V_a) as given below:

$$m_{\text{exper}} = m_s + m_t - \rho_g(V_s + V_a) \tag{3.59}$$

$$m_t = m_{\text{exper}} - m_s + \rho_g V_s + \rho_g V_a \tag{3.60}$$

to,

$$m_e = m_{\text{exper}} - m_s + \rho_g V_a \tag{3.61}$$

as

$$V_s = \frac{m_s}{\rho_s} \tag{3.62}$$

therefore,

$$m_e = m_{\text{exper}} - m_s + \rho_g\left(\frac{m_s}{\rho_s}\right) \tag{3.63}$$

and

$$m_e = m_t - m_s + m_t\left(\frac{\rho_g}{\rho_a}\right) \tag{3.64}$$

$$m_t = \frac{m_e}{\left(1 - \dfrac{\rho_g}{\rho_a}\right)} \tag{3.65}$$

$$m_t = \frac{m_{\text{exper}} - m_s + m_s\left(\dfrac{\rho_g}{\rho_s}\right)}{\left(1 - \dfrac{\rho_g}{\rho_a}\right)} \tag{3.66}$$

The adsorbent's ability to retain its reversible adsorption capacity during repeated gas charge and discharge cycles is referred to as long-term cycling stability. Factors affecting the cycling stability of a typical adsorbent can include porosity content and size, atomic and molecular structure, adsorption forces, physical and chemical degradation during cycles, decomposition and recomposition, operating pressure and temperature, gas-phase impurities and contaminants, the mobility of the adsorbent's structural units, substituent atoms in the lattice, the formation of dislocations during initial activation and extended cycling, vacancy concentrations, and the degree of phase separation.

3.2 PORE CLASSIFICATION IN POROUS MATERIALS

Classification of pores is a basic requirement for a comprehensive characterization of porous solids. The purpose of these classifications is to organize pores into classes by categorizing them based on characteristics like structure, size, accessibility, shape, and similar factors. The relative performance of porous substrates is highly dependent on the internal pore structure of each material. Therefore, properties of porous materials such as internal geometry, size, and connectivity are necessary for estimating their adsorption capacity. General pore classifications in solids, based on pore origin, structure, size, and accessibility with respect to origin and structure, include intraparticle pores, which are allocated to individual particles and defined as structurally intrinsic pores. The newly formed pores are classified as injected intrinsic pores. Extrinsic types of intra-pores are formed by a reaction in which a foreign substance is impregnated in the parent material and subsequently removed by the above modification procedures (pure type). Another type of extrinsic pores, known as pillared pores, is produced by substances such as metal hydroxides. In some cases, extrinsic intra-pores could also be inter-particle pores. Some porous materials are consolidated into relatively rigid, macroscopic structures whose dimensions exceed those of the pores by many orders of magnitude (agglomerates). Other materials that are not consolidated and consist of non-rigid, loosely packed assemblies of individual particles are referred to as aggregates. The particles may be non-porous and therefore surrounded by a network of inter-particle voids. These voids have properties that depend on the size, shape, and manner of packing of the constituent particles. The particles can be significantly porous, which means that a distinction exists between internal (or intraparticle) voids and inter-particle voids. Internal pores are smaller in size and total volume than the voids between particles (inter-aggregation (intra-cluster) and inter-cluster pores) [4]. Other pore classifications are associated with accessibility to the surroundings. The pores in direct contact with the external surfaces are called open pores; these are accessible to molecules or ions in the environment. The blind (dead-end) pores are open at one end or through both ends (through pores). Closed pores result from insufficient evolution of gaseous substances. Despite the fact that closed pores are not associated with gas adsorption and permeability, they influence the general properties of solid materials. Figure 3.13 represents a schematic of pore classification. From a standards point of view, closed pores cannot be penetrated by gas at ambient temperature. Similarly, a closed pore can be an open pore whose width/diameter is smaller than the size of the gas molecules (ultra-pores).

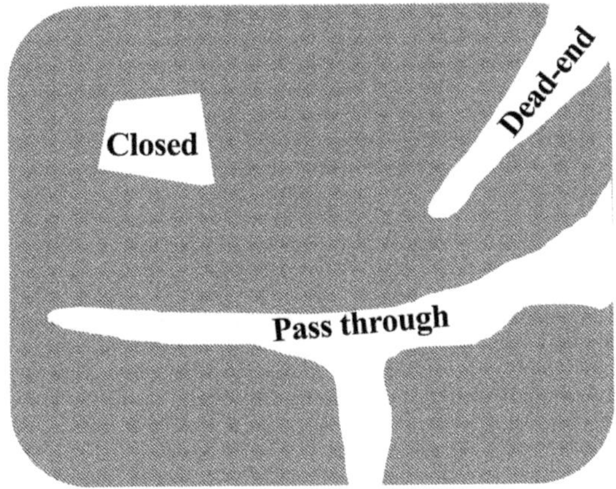

FIGURE 3.13 Pore classification: open and closed. The concept of pass through and dead-end pores are essential in gas adsorption of porous substrates. Created, designed, and introduced by the author.

TABLE 3.2
Size Classification of Pores in Adsorption Systems

Type	Range (nm)
Macro	>50.0
Meso	2.0–50.0
Micro	<2.0
Super-micro	0.7–2.0
Ultra-micro	<0.7
Sub-micro	<0.4

These are considered sub-categories of closed pores and chemically closed pores. In another classification, pores can be divided based on geometrical shapes such as cylinder, rhomboid, elliptical, square, slit-shaped (sheet-like in clays and activated carbon), cone-shaped (funnel-shaped), cylindrical (in activated oxides like alumina), prisms (fibrous structures), cavities and windows, slits (possible in clays and activated carbons), or spheres. Combinations of these shapes are often reported in real-life structures with specific geometry, pore size, orientation, location, and type of connectivity.

Porous materials with similar apparent porosity types but different pore sizes and geometries show various reactions to the same conditions. Pore size is a major property in the practical applications of porous materials. Pore size has a specific meaning when the geometrical shape of the pores is defined and known. It is mostly based on their width, the diameter of cylindrical pores, the distance between two sides of a slit-shaped pore, or the smallest dimension in a fissured pore. Table 3.2 reports the most common pore size classifications in ranges of macro-, meso-, and micropores.

According to function, the pores are divided into two general categories: transport pores, which constitute macropores with a radius > 25 nm, and adsorbing pores, which can be classified as mesopores (1–25 nm), micropores (0.2–1 nm), and sub-micropores (<0.2 nm). The effective pore radius is measured normal to the direction of movement of the molecule during adsorption by the pore. The radius itself can serve as a characteristic in cylindrical pores and as half-width in the case of the fissure pores. In relative pore size classification, macropores ($r > 1000$ nm) larger than individual element size; micropores ($100 < r < 1000$ nm) are pores that are similar in size to the structural elements; sub-micropores ($r < 100$ nm) are pores smaller than structural particles; and ultra-micropores ($r < 1$–2 nm) are pores found inside structured elements. The categories in large-scale pores include inter-cluster (macropores with 10^4–10^6 nm), inter-aggregation (micropores with 1–30×10^3 nm), inter-particle range (sub-micropores with 25–1000 nm), and intraparticle (ultra-micropores with <3–4 nm).

3.3 KINETICS OF ADSORPTION

The exposure of a solid adsorbent surface to a gas flow can result in the gas molecules hitting the solid surface. Some of these molecules will be adsorbed to the surface, while others will bounce back. For any system, the rate of adsorption is initially significant as the adsorbent surface is vacant. Nevertheless, as the surface becomes covered by the adsorbed molecules, the adsorption rate will decrease. Meanwhile, the rate of desorption (the rate of adsorbed molecules that bounce off the surface), which occurs from the covered adsorbent surface, is initially very significant. As time progresses, the desorption rate will decrease while the adsorption rate continues to increase until the rate of adsorption is equal to the rate of desorption, which indicates the equilibrium point. During this stage, it can be said that the solid is in adsorption equilibrium with the gas. The equilibrium is considered dynamic, as the number of molecules adsorbed at the surface and the number of molecules bouncing off the surface are equal. Nonetheless, if the flows of molecules leaving and attaching to the adsorbent surface are not balanced, the process will either be adsorption or desorption. It should be noted that there are no barriers to prevent atoms or molecules from approaching the surface and entering the physisorption well. Therefore, the process is not activation-dependent, and the kinetics of physisorption are always fast.

As adsorption is represented thermodynamically by isotherms, a close study of isotherms from a kinetics perspective can indicate system kinetics. The dynamic equilibrium in a Langmuir isotherm-controlled system (between gas adsorbed on a surface and molecules in the gas phase) makes it conceivable to derive an isotherm for the adsorption process by considering and equating these rates for the two processes. The adsorption and desorption rates can be calculated as follows:

$$R_{\text{ads}} = \frac{f(\theta)P}{\sqrt{2\pi mkT}} \exp\left(-\frac{E_a^{\text{ads}}}{RT}\right) \tag{3.67}$$

$$R_{\text{des}} = \upsilon f'(\theta) \exp\left(-\frac{E_a^{\text{ads}}}{RT}\right) \tag{3.68}$$

Balancing these two rates yields

$$\frac{Pf(\theta)}{f'(\theta)} = C(T) \tag{3.69}$$

where θ is the fraction of sites occupied at equilibrium and the terms $f(\theta)$ and $f'(\theta)$ contain the pre-exponential surface coverage dependence of the rates of adsorption and desorption, respectively. The other factors have been taken to the right side to give a temperature-dependent "constant" characteristic of this particular adsorption process. We now need to make certain simplifying assumptions regarding $C(T)$. The first step is one of the key assumptions of the Langmuir isotherm.

$$f(\theta) = c(1-\theta) \tag{3.70}$$

$$f'(\theta) = c'\theta \tag{3.71}$$

The forward adsorption process exhibits kinetics with a first-order dependence on the concentration of vacant surface sites and a first-order dependence on the concentration of gas particles (proportional to pressure). The reverse desorption process exhibits kinetics with a first-order dependence on the concentration of adsorbed molecules.

$$\frac{P(1-\theta)}{\theta} = B(T) \tag{3.72}$$

With

$$B(T) = \left(\frac{c'}{c}\right)C(T) \tag{3.73}$$

Using the Langmuir isotherm expression for surface coverage,

$$\theta = \frac{bP}{1+bP} \tag{3.74}$$

with

$$b = 1/B(T) \tag{3.75}$$

is a function of temperature and contains an exponential term:

$$b \propto \exp\left[\left(E_a^{\text{des}} - E_a^{\text{ads}}\right)/RT\right] = \exp\left[-\Delta H_{\text{ads}}/RT\right] \tag{3.76}$$

$$\Delta H_{\text{ads}} = E_a^{\text{des}} - E_a^{\text{ads}} \tag{3.77}$$

Therefore, b can be regarded as a coverage constant if the enthalpy of adsorption is independent of coverage, which is the second major assumption of the Langmuir isotherm.

Kinetic models are developed to describe experimental kinetic gas sorption and desorption data. Despite the fact that the kinetics of the adsorption and desorption of many molecular gases, like hydrogen, by microporous materials are swift, they also hold practical interest for system designs, although their experimental determination is challenging. Many reaction kinetics models are developed under the assumption that there exists a single rate-limiting step. In the case of absorption and desorption, there are a number of processes involved, each of which could potentially be rate-limiting. For instance, phenomena such as surface penetration, gas diffusion, and gas-phase formation should be considered. Therefore, some of these possible models for these processes should be taken into account. In most models of the gas absorption and desorption process, a spherical particle shape and the shrinking core model are presented with consideration for five partial reaction steps: (i) physisorption of molecular gas on the surface; (ii) dissociation of the molecular gas and subsequent atomic chemisorption; (iii) surface penetration by atomic gas; (iv) gas diffusion; and (v) complex compound formation. Models have been presented as a combination that incorporates the physisorption, chemisorption, surface penetration, and diffusion partial steps. This allows for the possibility of different partial reaction steps being rate-limiting at different stages of the absorption or desorption processes. For instance, a surface-related process could determine the initial absorption rate, with a subsequent transition to a diffusion-limited regime at longer times. The complex nature of the heterogeneous gas–solid reaction means that the laws of first-order reaction kinetics are not applicable to the description of the kinetics of absorption and desorption. These assumptions describe the case of a single-phase homogeneous reaction in which the reaction rate is proportional to the unreacted fraction, while others are often used to describe the kinetics of nucleation and growth processes. It is believed that the determination of the rate-limiting step through the analysis of sorption kinetics for the spherical particle model is achievable, but only under certain considerations. Considering the assumption that particle size distribution is possible, there are variations in the shape of the adsorption zones, and there exists distribution in the time at which the reaction proceeds for each of the zones. In most cases, a single rate-limiting step is inconceivable. During the evaluation of gas sorption kinetics, temperature variation due to the exothermic or endothermic nature of the absorption and desorption process should also be considered. The temperature variation is not homogeneous and can create thermal gradients. Overall, these temperature transients do not affect equilibrium uptake estimates, as evaluations are made after thermal equilibrium. Real-life surface conditions are also different from ideal conditions (a near-perfect surface in an ultra-high vacuum (UHV) gas purity) in models, including unclear polycrystal crystallographic (hkl) planes of adsorption on the surface and contamination in the gas phase. Experimental surfaces also contain oxides (oxidation due to oxygen exposure), hydroxides, and many other surface functional groups. As a result, the gas adsorption reaction may involve the reduction of the surface oxide layer, which affects absorption behavior. In most cases, a sigmoidal-type behavior (type IV and V sigmoidal adsorption isotherms) is observed, which is related to the successive activation of the sample surface that has been poisoned by impurities rather than to internal kinetics due to the nucleation and growth process. Extrinsic parameters such as surface passivation (to a minor extent) and the presence of gaseous impurities (at very low concentrations) can interfere with the adsorption and dissociation process. Besides these factors, the defective nature of real surfaces and deviations from the specified temperature range strongly affect the bulk kinetic uptake of the gas phase [5].

The surface penetration step is applied using an idealized surface barrier model, which has been utilized in various forms to describe the flux of the gaseous phase penetrating the surface in a number of combined kinetic estimations. Gas penetration through the surface (permeability) is described by different fluxes, including f_1 the flux from the gas phase onto the surface; f_2 the flux from the surface to the gas phase; f_3 the flux from the surface to the bulk; and f_4 the return flux from the bulk to the surface. f_1 is dependent on the flux of molecules hitting the surface (C), the sticking coefficient (S), and the number of substrate atoms per unit area (Ns). The flux hitting the surface can be calculated from kinetic theory as follows:

$$\Gamma = \frac{p}{\sqrt{2\pi m_G k_B T}} \tag{3.78}$$

where P is the hydrogen pressure, m_G is the mass of the gas, k_B is the Boltzmann constant, and T is the temperature. The estimations of second-order Langmuir kinetics lead to the dependence of the sticking coefficient S on the surface coverage θ as follows:

$$S = S_0 \left(1 - \theta\right)^2 \tag{3.79}$$

where S_0 is the sticking coefficient for the bare surface. Therefore, $f1$ and $f2$ are estimated as follows:

$$f_1 = \frac{2\Gamma S_0}{N_s}\left(1 - \theta\right)^2 \tag{3.80}$$

$$f_2 = -K\theta^2 \tag{3.81}$$

with the rate constant K defined as follows:

$$K = K_0 \exp\left(-\frac{2E_D}{RT}\right) \tag{3.82}$$

where E_D is the surface energy per hydrogen atom, R is the universal gas constant, and T is the temperature. f_3 can be calculated as follows:

$$f_3 = -v\theta\left(1 - x\right) \tag{3.83}$$

where x represents the atomic fraction of gas atoms in the bulk and the rate constant v representing the jump frequency from the surface to the bulk. It can be defined as follows:

$$v = v_0 \exp\left(-\frac{E_A}{RT}\right) \tag{3.84}$$

where E_A is the activation energy for a jump from the bulk to the surface. And f_4 is given as follows:

$$f_4 = \beta(1-\theta)x \tag{3.85}$$

where the rate constant β represents the jump frequency from the surface to the bulk and is expressed as follows:

$$\beta = \beta_0 \exp\left(-\frac{E_B}{RT}\right) \tag{3.86}$$

The term E_B is the activation energy for a jump from the surface to the bulk. By balancing these fluxes, the following expression and two differential equations can be derived:

$$\frac{d\theta}{dt} = \frac{2\Gamma s_0}{N_s}(1-\theta)^2 - K\theta^2 - v\theta(1-x) + \beta(1-\theta)x \tag{3.87}$$

$$\frac{dx}{dt} = \frac{v\theta}{N_l}(1-x) - \frac{\beta}{N_l}(1-\theta)x \tag{3.88}$$

where N_s is the number of substrate atoms per unit area and N_l is the number of substrate layers in bulk. x is assumed to be small compared to 1, and $(1-x)$ is neglectable. Additionally, x and θ are assumed to be in quasi-equilibrium after an initial and very rapid transient. Also, $d\theta/dt$ is assumed to be very small compared to $N_l \dfrac{dx}{dt}$. As a result, the correlations can be rewritten as follows:

$$\left(1+\frac{\beta x}{v}\right)^2 \frac{dx}{dt} = \frac{2\Gamma S_0}{N_s N_l} - \frac{K\beta^2 x^2}{N_l v^2} \tag{3.89}$$

with following variables defined to further improve the model's accuracy:

$$y = \frac{x}{x_{\max}} \tag{3.90}$$

$$a = \frac{2\Gamma S_0}{N_s N_l x_{\max}} \tag{3.91}$$

$$b = \left(\frac{\beta}{v}\right)x_{\max} = \frac{\theta_{\max}}{1-\theta_{\max}} \tag{3.92}$$

$$\frac{(1-b)^2}{2}\ln(1+y) + \frac{(1-b)^2}{2}\ln(1-y) - b^2 y = at \tag{3.93}$$

$$\frac{1}{y} - \frac{1}{y_0} + 2b\ln\left(\frac{y_0}{y}\right) + b^2(y_0 - y) = at \tag{3.94}$$

where y_0 is the value of y at the start of the desorption process.

The gas diffusion mechanism is defined by random molecular movement from areas of high partial pressure to areas of low partial pressure in the gaseous state and throughout a solid substrate. It can be described by the amount of gas that has entered and passed through a typical substrate powder (sphere) or thin foils (sheet/thin film). While the surface concentration is assumed to be constant, the mass change due to the diffusion of a substance in a substrate (a dimension) as a function of time (t) is given as follows:

$$\frac{M_t}{M_\infty} = 1 - \frac{6}{\pi^2} \sum_{n=1}^{\infty} \frac{1}{n^2} \exp\left(-\frac{Dn^2\pi^2 t}{a^2}\right) \tag{3.95}$$

where M_t is the amount of substance that has entered the sphere in a certain time t, M_∞ is the amount after infinite time, and D is the diffusion coefficient. The above expression is applicable for the conditions of an isobaric gas sorption measurement. For a plane thin film substrate, with the surface concentration constant on both sides, the mass change as a function of time is expressed as follows:

$$\frac{M_t}{M_\infty} = 1 - \sum_{n=0}^{\infty} \frac{8}{(2n+1)^2 \pi^2} \exp\left(-\frac{(2n+1)^2 \pi^2 t}{4l^2}\right) \tag{3.96}$$

where l is the thickness of the sheet.

The kinetics of solid-state phase transformations can be described by the transformation in terms of a time constant (k) and an exponent (n) as follows:

$$\alpha = 1 - \exp\left(-kt^n\right) \tag{3.97}$$

where α is the reacted fraction and t is the time, through random spatial nucleation and an exponential decrease in the nucleation rate. The value of n depends on the geometry (including the dimensions) of the reaction. It can also be applied in linear form as follows:

$$kt = \left[-\ln(1-\alpha)\right]^{1/n} \tag{3.98}$$

Gas atom diffusion between interstitial sites in the substrate lattice occurs through different mechanisms, where the dominant mechanism depends primarily on temperature. Below the typical threshold temperature, processes are primarily quantum mechanical (gas atoms simultaneously exhibit characteristics of both particles (tiny pieces of matter) and waves (a disturbance or variation that transfers energy)). At lower temperatures, the coherent tunneling mechanism dominates the diffusion phenomenon (scattering processes are negligible and transport occurs in the absence of scattering). As the temperature increases, phonon-assisted tunneling becomes the primary mechanism (the tunneling of a gas atom or consecutive holes through diffusion barriers is accompanied by the absorption or emission of quantized energy). As the temperature increases further, the gas atoms behave as classical mechanical particles that migrate via over-barrier jumps activated by thermal excitations. At the highest temperatures, the gas atoms behave like those in a free gas and undergo free motion through the lattice. The quantum behavior of gas atoms at high temperatures can be ignored for the lower temperature regimes, and they act as classical particles. The rate at which gas atoms diffuse through an adsorbent can be characterized by its diffusion coefficient, categorized as the chemical diffusion coefficient (D_c) and the tracer diffusion coefficient (D_t). D_c describes the bulk diffusion of particles or atoms in the presence of concentration gradients, and D_t describes the diffusion of individual and traceable particles. The D_c factor can be measured via bulk or macroscopic measurements, and the D_t value can be measured using microscopic measurement techniques such as quasi-elastic neutron scattering and nuclear magnetic resonance. The D_c value is more practical as it describes the motion of gas atoms through a substrate as the system is charged or discharged. The D_t term indicates the rate at which interstitial gas atoms can transport through the lattice. The relation between D_c and D_t can be calculated as follows:

$$D_t = \frac{D_c k_B T}{c \left(\dfrac{\delta \mu}{\delta c} \right)} \tag{3.99}$$

where k_B is the Boltzmann constant, T is the temperature, c is the gas concentration, and μ is the chemical potential. The chemical diffusion coefficient can be calculated by fitting data to a solution where the diffusion equation for gas uptake under isothermal isobaric conditions applies. The process is gas atom diffusion-limited.

REFERENCES

1. Broom DP (Ed.) *Hydrogen Storage Materials: The Characterisation of Their Storage Properties.* New York: Springer, 2011.
2. Keller JU, Staudt R. *Gas Adsorption Equilibria: Experimental Methods and Adsorptive Isotherms.* New York: Springer, 2005.
3. Varin RA, Czujko T, Wronski ZS. *Nanomaterials for Solid State Hydrogen Storage.* New York: Springer, 2009.
4. Bolis V. Fundamentals in adsorption at the solid-gas interface: Concepts and thermodynamics. In Auroux A (Ed.), *Calorimetry and Thermal Methods in Catalysis.* New York: Springer, 2013.
5. Rupali N, Sumita S, Hudson SL, Amaya SL, Tanna A, Sharma M, Achayalingam R, Sonkaria S, Khare V, Srinivasan SS. Recent developments in state-of-the-art hydrogen energy technologies: Review of hydrogen storage materials. *Solar Compass*, 2023, 5: 100033.

4 Hydrogen Adsorption

Scope

Storing hydrogen in atomic/molecular states in solid compounds and on solid substrates via adsorption is a feasible solution for stationary and on-board applications of hydrogen as an energy source. Hydrogen storage through adsorption includes confinement in extra small (most likely in the nanometer range) adsorption sites (i.e., pores) of high-surface-area materials (such as porous structures), shaping future prospects for applications like fuel cell-powered vehicles. Hydrogen adsorption is being considered one of the most promising storage methods for fulfilling the requirements of the US Department of Energy.

4.1 SOLID-STATE HYDROGEN STORAGE SYSTEMS

The most important criteria for selecting solid-state hydrogen storage systems to be adopted as storage systems include applicable thermodynamics, swift adsorption–desorption kinetics, large storage capacity (high volumetric and gravimetric density), resistance against the number of cycles for both adsorption and desorption, mechanical strength and durability, and proper heat transfer medium. Hydrogen can be stored physically in gas/liquid form or in/on solid-state materials. Gas-based storage requires high pressure (in the 350–700 bar range), and liquid storage requires cryogenic temperatures (as hydrogen T_b is about −252.8°C at 1 atm). Hydrogen can also be stored and accumulated by adsorption (on solid substrate materials) or by absorption (in solid compound materials) [1]. These options are developed to facilitate on-board vehicle applications, improve material handling and equipment, and portable power applications. Table 4.1 summarizes the application and minimum storage requirements. These requirements are critical for the development and planning of future hydrogen storage systems. All are aimed at commercializing domestic applications, especially hydrogen-fueled vehicle platforms, according to

---

TABLE 4.1

Minimum Requirements for Hydrogen Applications

	System gravimetric capacity (kWh/kg)	System volumetric capacity (kWh/L)
Domestic vehicles	1.5	1.0
Materials handling	–	1.7
Portable power medium ($\leq$2.5 W)	1.0–1.3	1.3–1.7
Portable power medium (>2.5—150 W)	1.0–1.3	1.3–1.7

DOI: 10.1201/9781003488828-4

consumer expectations for range, internal space, refueling/recharging time, and overall vehicle performance. The high-density hydrogen storage system (battery) poses a significant challenge for stationary, portable, and transportation power-generating applications [2].

The solid-state hydrogen storage systems are solutions to all the abovementioned challenges. Some material choices include organic polymers, metal–organic frameworks (MOFs), composites/hybrids, alloys, and metal hydrides, borohydrides, complex hydrides, metal oxides and mixed metal oxides (spinels), silicates such as clay and zeolites, and carbon materials (graphite, nanotubes, and graphene). Hydrogen storage can be categorized by various storage methods, including high-pressure gas cylinders, liquid hydrogen, physisorption of hydrogen on materials with a high specific surface area, hydrogen intercalation in solid-state materials such as metals and complex hydrides, and storage of hydrogen based on metals and water. Hydrogen storage systems are developed to pack hydrogen in a small volume (close-packing), enhance volumetric density, increase discharge capacity, and improve charge–discharge efficiency. The hydrogen sorption/desorption typical cycle is represented in Figure 4.1. From a thermodynamic point of view, compressing the gas, decreasing the temperature below the critical temperature, and/or reduced repulsion due to the interaction of hydrogen with other materials are the main solutions.

The hydrogen molecule contains a kinetic energy (E_k) which is related to the mean speed ($\bar{C}$) and can be calculated as follows:

$$\bar{C} = \sqrt{\frac{8RT}{\pi M}} \tag{4.1}$$

where T is the temperature, M is the molecular weight of the gaseous hydrogen molecule, and R is the gas constant. At ambient temperature, the mean speed

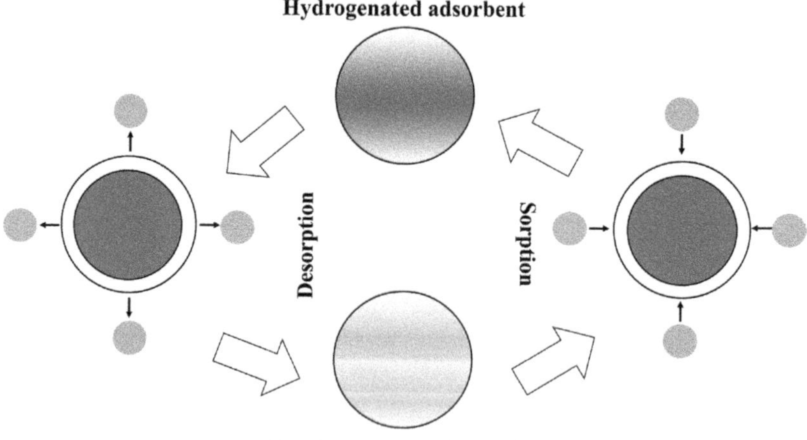

FIGURE 4.1 Typical hydrogen sorption/desorption cycle. Created, designed, and introduced by the author.

$\bar{C} = 1.77 \times 10^3$ m/s. The mean free path (λ), is the distance traveled by the hydrogen molecule between collisions under a typical very low pressure (p) of 1.3×10^{-6} Pa and at ambient temperature can be calculated as follows:

$$\lambda = \frac{kT}{\sqrt{2}\sigma p} \tag{4.2}$$

with the collision cross section of $\sigma \approx 0.3 \times 10^{-18}$ m^2 for hydrogen and k being the Boltzmann constant. At ambient temperature, the λ factor is estimated to be 8170 m. The proton drift speed (s) can be calculated as follows:

$$s = uE \tag{4.3}$$

with the mobility (u) of protons in water at ambient temperature estimated as 36.23×10^{-8} m^2 s^{-1} V^{-1}. With a typical potential difference between two electrodes separated by a distance of (d) and the applied electric field (E), the drift speed in the bulk electrolyte can be estimated for a hydrated proton. However, the electric field (E) in the electric double layer (within the electrode's reach) is smaller than the typical drift speed. Although the mobility value in the electrode's immediate zone is significantly higher than in the bulk electrolyte, it is still between 50 and 500 times lower than the mean speed of hydrogen molecules in the gas phase at ambient temperature. These physical parameters indicate that the behavior of the two precursors of adsorbed H (the hydrogen molecule in the case of gas-phase adsorption and the hydrated H$^+$) is different. The transport of protons can be considered as diffusion H$_9$O$_4^+$, contrary to the established theory that proton transport can be expressed by a series of hops between water molecules. The kinetic energy possessed by hydrated protons is insufficient to induce any mechanical damage to the electrode surface upon contact and discharge. Nevertheless, the kinetic energy of hydrogen molecules in the gas phase induces surface damage through sputtering. These two conditions include differences such as (i) the absence or presence of solvated ions and solvent molecules near the free surface and their effect on structure-forming or structure-breaking phenomena; (ii) the formation of new compounds at the surface of the electrode as a result of electrodeposition, electrooxidation, and metal–solvent interactions; (iii) the absence/presence of a strong electric field across the interphase and its role in the space charge effect; (iv) charge transfer processes at the interface and related structural changes; and (v) fast exchange of hydrogen molecules across the metal–gas interphase under gas-phase conditions.

4.2 PHYSISORPTION OF HYDROGEN

Theoretical aspects of general gas adsorption to solid surfaces are mentioned in detail in the previous chapter. Physical adsorption of gas or vapor (adsorbate) on solid surfaces (adsorbent) involves the accumulation of molecules at the interface. This interaction phenomenon is governed by attractive van der Waals or London dispersion forces and originates in the attraction between fluctuating and induced dipoles of gas-phase molecules and atoms on the solid surface (resonant fluctuations of charge distributions as dispersive interactions). Hydrogen adsorption on microporous structures occurs through van der Waals' interactions between non-dissociated hydrogen molecules and active surface sites. In the hydrogen physisorption process, H_2 molecule interacts with several solid surface atoms. This interaction consists of an attractive component that decreases with the distance between the molecule and the surface to the power of 6 and a repulsive term that decreases with distance to the power of 12. Consequently, the potential energy of the molecule shows a minimum at a distance of one molecular radius of the adsorbate (53–120 pm for the hydrogen atom–molecule). As the binding forces are primarily distance-dependent, comparatively weak, and therefore more susceptible to disturbance, the physisorption process is swift. Hydrogen can be reversibly stored (adsorbed) by physisorption, without any energy barrier between the gas phase and the adsorbed state, thus allowing for fast adsorption and desorption kinetics. Since the accessible surface area is the dominating factor in this phenomenon, highly porous materials with large internal specific surface areas are ideal for physisorption as a hydrogen storage mechanism. Furthermore, due to the low heat of adsorption (the intensity of binding hydrogen to the adsorbent surface) involved in the physisorption of hydrogen (ΔH_{ads}), adsorption is feasible without additional thermal energy systems and heat management apparatuses. ΔH_{ads} is the main characteristic of an adsorbent, providing significant information about the adsorbed hydrogen at ambient conditions. Generally, the low adsorption enthalpy involved in molecular adsorption on porous materials makes low temperatures essential for achieving relevant storage densities. The potential energy minimum of hydrogen adsorption is between 0.01 and 0.1 eV (1 and 10 kJ/mol). Due to weak interactions, significant physisorption can be observed at low temperatures. After the monolayers of hydrogen molecules are formed along the free surface, the H_2 molecules interact with the surface of the solid adsorbate. Consequently, the binding energy of the second layer of hydrogen molecules is similar to the latent heat of sublimation or vaporization of the adsorbate. As a result, the adsorption of hydrogen at a temperature equal to or higher than the boiling point at a certain pressure leads to the adsorption of a single monolayer. To measure the amount of adsorbate in a monolayer, the density of the adsorbate and the volume of the molecule provide details on the monolayer, including the minimum surface area S_{mol} for one mole of adsorbate on a substrate. For instance, in a close-packed, face-centered cubic -structured adsorbent, S_{mol} can be calculated from the molecular mass of the adsorbate (M_{ads}) and its density as follows:

$$S_{mol} = \frac{\sqrt{3}}{2}\left(\sqrt{2N_A} \cdot \frac{M_{ads}}{\rho}\right)^{2/3} \tag{4.4}$$

where N_A is the Avogadro constant (6.02×10^{23} mol^{-1}) [3].

According to these calculations, the monolayer surface area for hydrogen can be estimated as S_{mol} (H_2) ≈ 85920 m^2/mol. The amount of adsorbate M_{ads} on a substrate material with a specific surface area (S_s) is given by $M_{ads} = M_{ads} \cdot S_s / S_{mol}$. By choosing a certain solid adsorbate surface, such as carbon (a single-sided graphene sheet) with hydrogen adsorbent, the S_s can be estimated at approximately 1316 m^2/g with maximum adsorbed hydrogen as $M_{ads} \approx 3$. Through this theoretical approximation approach, one may conclude that the amount of adsorbed hydrogen is proportional to the specific surface area of the adsorbent [4]. Under gas-phase atmospheric conditions, the adsorption routes and the nature of the adsorbed hydrogen-containing compounds rely on electronic properties, physical surface structure, chemical oxidation state, and the presence or absence of hydrogen and oxides. It can also involve characteristics of the metallic surface and the operational temperature. At very low temperatures, gaseous hydrogen molecules approach the surface ($M(s)$) and experience interactions that lead to their physisorption:

$$M(s) + H_2(g) \xrightarrow{\text{Temperature} < 20\,\text{K}} M - H_{2,\text{Physiorption}}$$

The bond energy between $M(s)$ and the physisorbed hydrogen molecule is expressed by $E^\circ\left(M - H_{2,\text{Physiorption}}\right)$ and can be related to the standard enthalpy of physisorption $\Delta_{\text{physiorption}}E^\circ(H_2)$ with weak physical interactions through the following equation:

$$E^\circ\left(M - H_{2,\text{Physiorption}}\right) = -\Delta_{\text{physiorption}}E^\circ(H_2) \tag{4.5}$$

The main characteristic of the physisorption process is the intact nature of the hydrogen molecule's covalent bond without dissociation. The standard dissociation energy of the adsorbed hydrogen molecule ($D^\circ\,H_2$) and H–H bond length are slightly different from those of isolated hydrogen molecules. Although physisorption is an exothermic process $\left(\Delta_{\text{physiorption}}E^\circ(H_2) < 0\right)$, the amount of energy released is significantly less than the endothermic enthalpy required to dissociate the hydrogen molecule. Hydrogen electro-adsorption is the electrochemical process that can be accomplished in either acidic or basic aqueous solutions, as well as in non-aqueous solutions with dissolved H-containing acids. These acids serve as the proton source, or the solvent can dissociate and lead to proton formation. A proton (H^+) does not exist alone in an aqueous acidic solution; it rapidly combines with non-bonding electron pairs of water molecules, forming H_3O^+ with a strong chemical bond ($H_2O - H^+$). The enthalpy of formation of this compound $\left(\Delta H_f^\circ\left(H_3O^+\right)\right)$ is large and negative. This H_3O^+ proton attracts more H_2O molecules through a strong electric field, which leads to an additional rigid hydration shell and the formation of $H_9O_4^+$ cations. The formation of three more $H_3O^+ - H_2O$ bonds creates additional stabilization enthalpy. The term used includes ion–solvent dipole interactions as well as long-range order polarization of H_2O in the diffuse hydration shell; the rigid and diffuse hydration shells constitute an ionic atmosphere that surrounds the proton. The external electric field drives the transport of cations and anions within the electrolyte and creates a potential difference across the anode/electrolyte and cathode/electrolyte interphases. As hydrated H_3O^+ approaches the electrode double layer, a discharge occurs with the formation of electro-adsorbed hydrogen based on the single- electrode process as follows:

$$M(s) + H_3O^+ + e^- \xrightarrow{\text{Energy}} M - H_{\text{ads}} \tag{4.6}$$

where E is the potential energy at which the electro-adsorption of hydrogen occurs. There are different types of electro-adsorbed hydrogen, including under-potential deposited (H_U) and the overpotential deposited (H_O). H_U can become electro-adsorbed at potential levels of $E > E^{\circ}_{H^+/H_2}$ and H_O at potential levels of $E < E^{\circ}_{H^+/H_2}$ for $E^{\circ}_{H^+/H_2} = 0.0$ V as the standard potential of H^+/H_2 redox couple. Therefore,

$$M(s) + H_3O^+ + e^- \xrightarrow{\text{Energy}>0} M - H_U + H_2O \tag{4.7}$$

$$M(s) + H_3O^+ + e^- \xrightarrow{\text{Energy}<0} M - H_O + H_2O \tag{4.8}$$

The water molecules that form the hydration shell of H_3O^+ can be excluded, as the hydration shell surrounding the proton is disassembled upon charge transfer between the metal surface and H_3O^+. However, the water molecules freed in the process can interact with other soluble species in the electrolyte solution and react with the metallic electrode in the electric double layer. The H_O act as an intermediate and H_U acts and does not participate in the process. The electrolytic hydrogen gas generation involves three stages as follows:

$$M(s) + H_3O^+ + e^- \xrightarrow{\text{Energy}<0} M - H_O + H_2O$$

$$M(s) - H_O + H_3O^+ + e^- \xrightarrow{\text{Energy}<0} M(s) + H_2(g) + H_2O$$

$$2M(s) - H_O \xrightarrow{\text{Energy}<0} 2M(S) + + H_2(g) \tag{4.9}$$

The order of these reactions is referred to as the Volmer–Heyrovsky mechanism and the Volmer–Tafel mechanism. H_U and H_O can engage in interfacial transfer from the adsorbed state to a subsurface state (subsurface hydrogen or H_s and transfer from the subsurface state back to the adsorbed state hydrogen (H_{abs}) in metals and metallic alloys that are capable of absorbing hydrogen.

$$H_U \xrightarrow{\text{Energy}>0} H_s \rightarrow H_{abs}$$

$$H_O \xrightarrow{\text{Energy}<0} H_s \rightarrow H_{abs} \tag{4.10}$$

As the higher heat of adsorption causes stronger hydrogen interactions with the adsorbent and requires less pressure or cooling to reach the same amount of adsorbed hydrogen, the differential molar isosteric heats of adsorption (ΔH_{iso}) can also be estimated as the difference between the molar enthalpy of the gas phase (H_g) and the differential molar enthalpy of the adsorption (H_{ads}) system as follows:

$$H_{ads} = \left(\frac{\delta H}{\delta n} \right)_T$$

$$\Delta H_{iso} = H_g - H_{ads} \tag{4.11}$$

The heat of adsorption can be calculated by considering the non-ideal character of the gas phase at high pressures and the non-inert character of the adsorbent during adsorption:

$$\Delta H_{\text{iso}} = -RZ\left(\delta(\ln P)\Big/\delta\left(\frac{1}{T}\right)\right)_{n} \cdot \left(\left(1-\left(\frac{\delta V_n}{\delta n}\right)_T\Big/V_g\right)-\left(\delta P/\delta n\right)_T\right)\cdot\left(V_n - T\left(\delta V_n/\delta T\right)_n\right)$$

$$(4.12)$$

with R as the universal gas constant, $Z = P\cdot V_g/(RT)$ as the compressibility of an equilibrium gas phase at pressure P and temperature T, V_g as the specific gas-phase volume, $V_n = V_0\,(P,T)/m_0$ as the reduced volume of adsorbent/adsorbate system, and V_0 m_0 are the volume and mass of the adsorbent. It considers isothermal adsorption-induced deformation $\left(\dfrac{\delta V_n}{\delta n}\right)_T$, temperature isosteric deformation $\left(\dfrac{\delta V_n}{\delta T}\right)_n$, the isotherm of adsorption slope $\left(\dfrac{\delta \ln P}{\delta(1/T)}\right)_n$, and the non-ideal gas phase with compressibility Z.

According to estimations, adsorption of hydrogen in the range above the critical temperature does not significantly affect the relative adsorption deformation of the adsorbents (less than 1%), and the temperature-induced deformations of the adsorbents are negligible. Nevertheless, the differential molar volume of the adsorption system $\left(\dfrac{\delta V_n}{\delta n}\right)_T$ is significantly smaller than V_g. As a result, the heat of adsorption of hydrogen can be calculated as follows:

$$\Delta H_{\text{iso}} = -RZ\left(\delta(\ln P)\Big/\delta\left(\frac{1}{T}\right)\right)_{n} - \left(\delta P/\delta n\right)_T\cdot V_n \qquad (4.13)$$

Therefore, the value of ΔH_{iso} for hydrogen adsorption systems should not be estimated from the experimental isosteres of hydrogen adsorption. The graphical presentation of the abovementioned equation is shown in Figure 4.2. The decreasing trend with a decreasing slope is the primary specification of the relation. A higher difference between the densities of the adsorbed hydrogen gas and the bulk media indicates a more efficient hydrogen storage process. Nevertheless, due to the low critical temperature of hydrogen, approaching liquid densities in the adsorbed phase at ambient temperature is very difficult. When operating at moderate pressures, storage capacities do not dramatically exceed those of simple compression at the same pressure.

The adsorbent can be described by a linear function, as the linearity of adsorption isosteres is characteristic of many adsorption systems with a direct effect from the amounts of adsorbed hydrogen (n). Figure 4.3 shows the schematic isosteres plots for adsorbed hydrogen with different amounts of adsorbed hydrogen.

Consequently, for enhanced volumetric and gravimetric adsorption capacities of hydrogen in porous materials, operating at cryogenic temperatures drastically

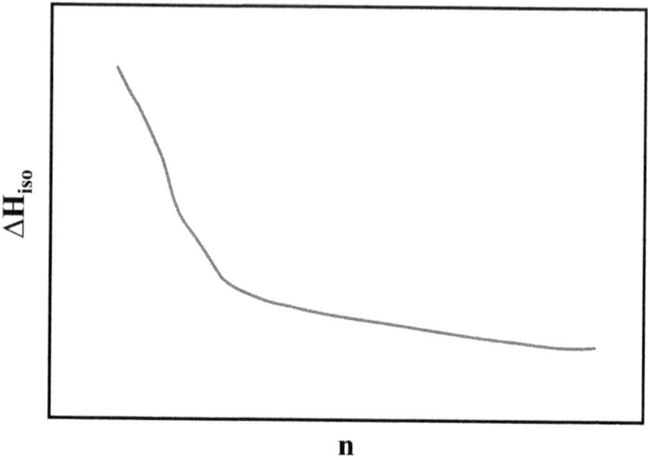

FIGURE 4.2 Graphical representation of isosteres for a typical hydrogen adsorbent. Created, designed, and introduced by the author.

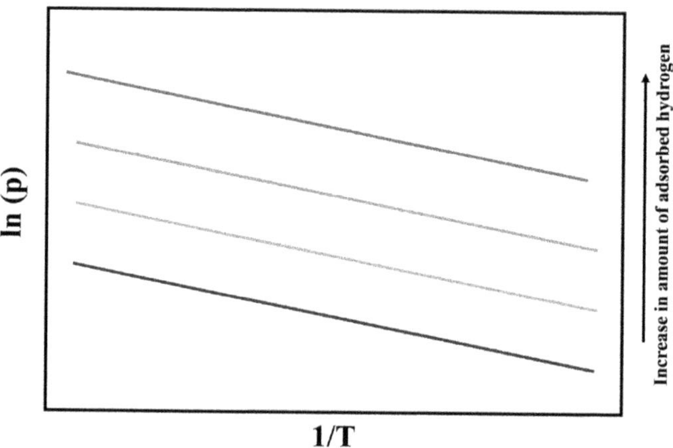

FIGURE 4.3 A schematic representation of isosteres plots for adsorbed hydrogen with different adsorption contents. Created, designed, and introduced by the author.

decreases the required pressure. Increasing the bulk density of the adsorbent (such as activated carbon compared to MOFs) provides a larger mass of adsorbent, which improves adsorption capacity in an isochoric system. The design and characterization of porous materials, along with their assessment as high-capacity adsorbents for hydrogen adsorption–desorption, are based on these conditions. If the overall energy consumption of a typical storage system can be primarily assessed by analyzing the thermodynamic routes of the process between different equilibrium states, more operational systems can be introduced. A more accurate assessment of the heat and power energy requirements indicates the dynamics of the system. The transient variations in operating parameters include pressure, temperature, and amounts of accumulated hydrogen during the charge and discharge steps. The dynamic adsorption process during hydrogen loading and discharge requires consideration of mass and energy balances, combined with an assessment of state to describe the gas-phase behavior and temperature-dependent equilibrium adsorption isotherms. Assuming the hydrogen gas phase acts as ideal is considered reasonable if the operating adsorption pressure remains moderate. In the pressure range of ~150 bar, the ideal behavior assumption for hydrogen gas is applicable, while in the case of higher working pressures (~700 bar) and cryogenic temperatures, the non-ideal behavior of the hydrogen gas phase should be considered, which requires implementing a real gas state. To describe the adsorption equilibrium of hydrogen onto microporous adsorbents (i.e., activated carbons and MOFs), modified Dubinin–Astakhov isotherms (high pressure and super-critical temperature) are more applicable. It indicates that

$$\Delta H_{\text{iso}} = -\Delta H_{\text{ads}} = \alpha \sqrt{-\ln\left(\frac{n_{\text{abs}}}{n_{\text{max}}}\right)} \tag{4.14}$$

with n_{abs} as the absolute amount of adsorbed hydrogen, n_{max} as the amount of adsorbed hydrogen corresponding to the saturation of the total available porous volume, and α as the enthalpy factor. The non-linear fitting of the isotherm (compared to Langmuir) was carried out by considering the density of the adsorbed phase as equal to the density of liquid hydrogen. The $\frac{n_{\text{abs}}}{n_{\text{max}}}$ term is the hydrogen fractional filling of pore volume, θ [5]. The excess adsorbed hydrogen is related to n_{abs} and the volume of the adsorbed phase (V_a) as given below:

$$n_{\text{exc}} = n_{\text{abs}} - \rho_g V_a \tag{4.15}$$

The typical excess hydrogen adsorption–desorption isotherms at different pressures can be represented as Figure 4.4.

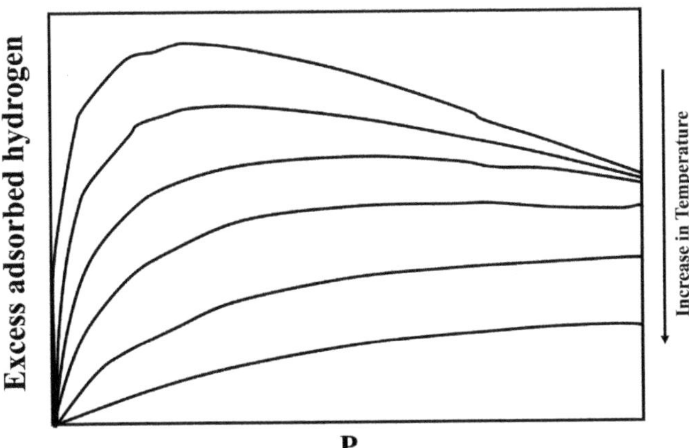

FIGURE 4.4 A schematic representation of excess hydrogen adsorption–desorption isotherms. Created, designed, and introduced by the author.

The selection of a proper temperature-dependent isotherm equation is therefore crucial for obtaining good predictive ability in process simulation. Additionally, by either considering ideal or non-ideal behavior for hydrogen gas and describing adsorption equilibrium data according to different forms of isotherm equations, the spatial heterogeneities of the system may or may not be assumed. Estimations can be designed using lumped measurements (where different values of adsorption are combined into a single value) or distributed measurements (where the spatial variation of inputs and parameters is represented). For instance, in zero-dimensional (0D) systems, lumped estimations are used where the variables of the system are assumed to be uniform throughout the entire adsorbent volume, allowing for overall pressure, temperature, and hydrogen amounts to be computed at each time step for the whole system. In such systems, the mass and energy balance equations can be calculated as follows:

$$m_{\text{in}} - m_{\text{out}} = V_{\text{system}} \left(\Phi \cdot \frac{\delta \rho_g}{\delta t} + \rho_s M \frac{\delta n_n^{\text{ads}}}{\delta t} \right) \tag{4.16}$$

$$H_{\text{in}} - H_{\text{out}} + H_{\text{ads}} = V_{\text{system}} T \left(\Phi \cdot \frac{\delta \rho_g U_g}{\delta t} + \rho_s M \frac{\delta \left(n_n^{\text{ads}} U_n \right)}{\delta t} + \rho_s M \frac{\delta U_s^{\text{ads}}}{\delta t} + \rho_w \frac{\delta U_w^{\text{ads}}}{\delta t} \right)$$

$$\tag{4.17}$$

Similarly, in one-dimensional (1D) systems,

$$\Phi \cdot \frac{\delta \rho_g}{\delta t} + \rho_s M \frac{\delta n_n^{\text{ads}}}{\delta t} = - \frac{\delta \left(V \cdot \rho_g \right)}{\delta x} \tag{4.18}$$

$$- \frac{\delta}{\delta x} \left(V \cdot \rho_g H_g T \right) - \frac{\delta}{\delta x} \left(V \cdot \rho_g \cdot n_n^{\text{ads}} \cdot H_{\text{ads}} \right)$$

$$= T \left(\frac{\delta \left(\rho_g U_g \right)}{\delta t} + \rho_s M \frac{\delta \left(n_n^{\text{ads}} U_n \right)}{\delta t} + \rho_s \frac{\delta U_s^{\text{ads}}}{\delta t} + \rho_w \frac{\delta U_w^{\text{ads}}}{\delta t} \right)$$

and for one-dimensional (1-D) system,

$$\Phi \cdot \frac{\delta \rho_g}{\delta t} + \rho_s M \frac{\delta n_n^{\text{ads}}}{\delta t} = - \frac{\delta}{\delta x} \left(V \cdot \rho_g \right) - \frac{1}{r} \frac{\delta}{\delta r} \left(r V \rho_g \right) \tag{4.19}$$

The multi-dimensional approximations allow for system evaluation in the axial direction. The corresponding mass and energy calculations are established according to the volumes of the system. The significant advantage of multi-dimensional approximations is their accuracy due to the local adjustment of parameters related to mass and heat transfer kinetics, such as intra-particle hydrogen diffusivity or heat transfer coefficients. Considering the compromises obtained from the numerical simplification of zero-dimensional estimations, the limited number of adjusted lumped parameters helps in obtaining time-dependent profiles of pressure, temperature, and hydrogen storage capacities.

Hydrogen desorption equilibria describe the amount of hydrogen adsorbed at equilibrium under given pressure and temperature conditions. In this respect, it acts as a combination of simulating hydrogen adsorption systems and isotherm models. Essentially, it is crucial for the proper evaluation of the storage capacities of adsorption systems. The characteristic curve (developed based on the Dubinin equation modified according to the potential theory developed by Polányi) expresses the relationship between the state of compression of an adsorbed gas (i.e., hydrogen) and the forces facilitating the interaction of the adsorbent's surface, known as the adsorption potential. For a combination of given adsorbate–adsorbents, a single curve independent of temperature describes the volume of the adsorbate in the molar adsorbed phase (V_n^{ads}) as a function of the adsorption potential. The fraction of the micropore filling volume occupied by the adsorbed phase, according to the functional form of the Weibull distribution, can be described as follows:

$$V_n^{ads} = V_{sat}^{ads} \exp\left(-\left(\frac{A}{\varepsilon}\right)^m\right) \tag{4.20}$$

where V_n^{ads} is the maximum molar volume that the adsorbate can occupy, estimated as the total volume of the micropores; ε is the characteristic energy representative of the adsorbent–adsorbate system; and m is a constant that expresses the pore heterogeneities. The *m* factor is estimated to be less than or equal to 2. It can increase to the 3–6 range for microporous carbon molecular sieves (CMS). The differential molar work of adsorption (A) can be estimated as follows:

$$A = RT \ln\left(\frac{f_s\left(T, P_s\left(T\right)\right)}{f\left(T, P\right)}\right) \tag{4.21}$$

where P_s is the saturated vapor pressure of the hydrogen adsorbate, and f and f_s are the adsorbate fugacities at temperature T. By assuming that for hydrogen adsorbate at the same fraction of the micropore filling volume, the ratio A over ε is constant and can be defined as the similarity coefficient:

$$\beta = \frac{\varepsilon}{\varepsilon^0} = \frac{A}{A^0} \tag{4.22}$$

where ε^0 and A^0 are, respectively, the characteristic energy and adsorption potential of the reference gas compared to hydrogen. The affinity coefficient β factor can be estimated through a variety of correlations, such as the ratio of the parachors of the hydrogen adsorbate to reference molecules: Π, Π^0 ($cm^3\ g^{1/4} s^{-1/2} mol^{-1}$):

$$\beta = \frac{\varepsilon}{\varepsilon^0} = \frac{\Pi}{\Pi^0} \tag{4.23}$$

For hydrogen, β is estimated to be 0.165. ε^0 is a reverse function of the micropore half-width x (nm) and can be estimated as follows:

$$\varepsilon^0 = \frac{13}{x} \tag{4.24}$$

with characteristic free energy calculated as follows:

$$\varepsilon = \varepsilon^0 \cdot \beta = 0.165\frac{13}{x} \tag{4.25}$$

By indicating the amount of fractional molar micropore volume participating in adsorption (V_n^{ads}), the molar adsorption capacity (n_n^{ads}) can be estimated from the density of the adsorbed phase as follows:

$$n_n^{ads} = V_n^{ads}\frac{\rho_n}{M} \tag{4.26}$$

where M is the molar mass of the hydrogen adsorbate. At boiling temperature or below, ρ_n can be estimated to be equal to the density of the bulk liquid. For the range of temperatures from the boiling point to the critical temperature, the density of the hydrogen-adsorbed phase is determined as a function of the thermal coefficient of limiting adsorption α and can be calculated according to the Dubinin–Nikolaev equation as follows:

$$\rho_n = \rho_b \exp\left(-\alpha\left(T - T_b\right)\right) \tag{4.27}$$

where ρ_b is the density of the hydrogen at boiling temperature T_b (K). The thermal coefficient α is constant and can be derived from the ratio of the density of the bulk liquid hydrogen at boiling temperature ρ_b to the density of hydrogen at critical temperature ($T_{critical}$) according to the following equation:

$$\alpha = \frac{\left(\dfrac{\rho_b}{\rho_{critical}}\right)}{\left(T_{critical} - T_b\right)} \tag{4.28}$$

As the density of hydrogen adsorbate at critical temperature relates to maximum compression, the $\rho_{critical}$ value can be derived from the van der Waals constant (b) as follows:

$$\rho_{critical} = \frac{M}{1000b} \tag{4.29}$$

with

$$b = \frac{1}{8} \cdot \frac{R \cdot T_{critical}}{P_{critical}} \tag{4.30}$$

which for hydrogen is about 0.026 l/mol. At temperatures higher than $T_{critical}$, the density of the adsorbed phase does not depend on temperature and is estimated as $\rho_{critical}$. The fugacities for hydrogen adsorbate in the gaseous phase at a certain temperature and pressure can be estimated as follows:

$$f = P \cdot \exp\left(\int_0^P\left(\frac{z(T,P)-1}{P}\right)\right)dp \tag{4.31}$$

where z is the compressibility factor of the gaseous hydrogen adsorbate. The hydrogen fugacity at saturation (f_s) is dependent on temperature and pressure, as well as whether it is below or above the critical condition. By replacing P with $P_s(T)$, the saturation pressure can be estimated as follows:

$$\ln(P_s) = K - N \cdot \frac{1}{T} \tag{4.32}$$

where coefficients K and N are estimated according to the critical pressure and temperature of hydrogen adsorbate and its normal boiling point at 1 atm. The above-mentioned equations can be used to describe the hydrogen adsorption process onto microporous adsorbents under super-critical conditions. When T is about $T_{critical}$, the linearity of adsorption isotherms during the transition from sub-critical to super-critical conditions relies on the validity outside the super-critical temperature range. However, for temperatures largely above the critical point, the assessment of fs becomes difficult, and the thermal invariance of both the characteristic energy ε and the non-homogeneity parameter m should not be considered. Assuming a linear temperature dependence of the characteristic energy of adsorption ε, the enthalpy factor a (J/mol) and entropy factor B (J/mol/K), the equation changes as follows:

$$n_n^{ads} = n_s^{ads} \exp\left(-\left(\frac{A}{a + BT}\right)^n\right) \tag{4.33}$$

where n_n^{ads} (mol/kg) is the absolute molar adsorption capacity in equilibrium in the gaseous phase, and n_s^{ads} (mol/kg) is the maximum molar adsorption capacity adsorbed at saturation of the micropore volume. The density of the adsorbed hydrogen phase can be assumed constant along the micropore volume filling until saturation. This equation is used to describe adsorption–desorption capacities in hydrogen storage reservoirs of hydrogen adsorption isotherms between boiling and ambient temperatures. As a proven correlation of isotherm parameters and textural properties of adsorbents for hydrogen adsorption, the accessible micropore volume V_p and the limiting adsorption capacity n_n^{ads} are linearly correlated with the specific surface area, while A and B coefficients are related to the average micropore size and total pore volume. The saturation pressure P_s can be considered as a fitting constant parameter to analyze the adsorption potential A, as it tends to be lower for adsorbents with smaller micropores.

In accordance with the fundamentals of the potential theory, the variations of the adsorbed phase volume are a function of the number of moles adsorbed hydrogen molecules and the density of the adsorbate as follows:

$$V_n^{ads} = \frac{M \cdot n_n^{ads}}{\rho_n} \tag{4.34}$$

The Dubinin–Astakhov model takes into account the temperature dependence of the molar adsorption capacity as follows:

$$n_n^{ads} = \left(V_s^{ads} \cdot \frac{\rho_n}{M} \cdot \exp\left(-\left(\frac{R \cdot T}{a + B \cdot T}\right)^m\right) \cdot \ln^m\left(\frac{f_s}{P}\right)\right) \tag{4.35}$$

The characteristic free energy is estimated from the micropore width x, with the term B considered as a best fit parameter as follows:

$$\varepsilon = a + BT = 0.165\frac{13}{x} + BT \tag{4.36}$$

4.3 CHEMISORPTION OF HYDROGEN

Chemisorption, a form of adsorption in which the adsorbed material is bonded together by chemical bond forces, is also proportional to surface area. Chemisorption has high specificity and occurs when there are chemical bonds between the hydrogen adsorbent and the adsorbate. For instance, chemisorption on metal hydrides is well established, where the absorption process involves the dissociation of hydrogen molecules and the linking of hydrogen atoms to the metal by strong covalent bonds. During chemical adsorption, metal hydrides are employed as a form of solid-state storage for hydrogen gas using a bond approach. Chemisorption includes the transfer, exchange, or sharing of electrons between nanoparticles (scaffolds) and the molecules due to the formation of covalent or ionic bonds. As a result, it has high specificity and is irreversible. This method involves stability, high storage density, and small space occupancy. Metal alkaline and earth alkaline hydrogen storage materials include Na and Li aluminum/boron hydride and alane. Magnesium hydride has significant benefits, including large storage capacity and lightweight properties [6]. The limitations of chemisorption include the need for high temperatures for desorption, as hydrogen adsorption–desorption is irreversible due to its strong bonding forces compared to physical adsorption, which is a simpler immobilization method with molecules attached to the surface of the adsorbent's scaffold (porous) through weak van der Waals' bonds, hydrogen bonding, or hydrophobic interactions. The near-surface mechanisms leading to chemisorption are shown schematically in Figure 4.5, including surface diffusion, volume diffusion, dissociation, surface sticking, and surface trapping.

During chemisorption, gaseous hydrogen molecules approach the metallic surface at temperatures higher than those characteristic of physisorption, resulting in a strong chemical interaction. The interaction between the hydrogen molecule and the

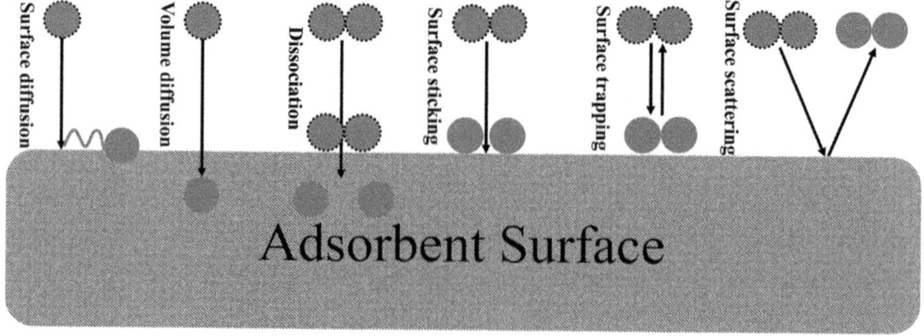

FIGURE 4.5 A schematic representation of near-surface mechanisms involved in chemisorption. Created, designed, and introduced by the author.

metallic surface leads to its dissociation into atomic H and an almost instantaneous formation of a chemical bond with the substrate as follows:

$$H_2 \rightarrow 2H^* \tag{4.37}$$

$$2M(s) + 2H^* \rightarrow 2M - H_{\text{chem}}$$

where H^* is the short-lived energetically rich H that develops a chemical bond with the metallic surface. The energy (enthalpy) increase from the formation of $2M - H_{\text{chem}}$ bonds $\left(E^0\left(M - H_{\text{chem}}\right)\right)$ is in the hundreds kj/mol range (for d-block metals), which is greater than the enthalpy input necessary for hydrogen molecule dissociation, and the extra energy will be released as the energy of chemisorption $\left(\Delta_{\text{chem}}^{H^0}\left(H_{\text{chem}}\right)\right)$, which can be calculated as follows:

$$E^0\left(M - H_{\text{chem}}\right) = 1/2 D^0\left(H_2\right) - \Delta H^0{}_{\text{chem}}\left(H_{\text{chem}}\right) \tag{4.38}$$

The $\Delta_{\text{chem}}^{H^0}\left(H_{\text{chem}}\right)$ is related to one mole of chemisorbed hydrogen atoms, which creates two mols of H_{chem} adsorbed atoms. While the H^* high-energy unit approaches the surface, the 1s orbital overlaps with the delocalized orbitals of the surface material. This can occur through activated and non-activated routes, where the hydrogen molecule dissociates with an energy release equal to two times $\Delta_{\text{chem}}^{H^0}\left(H_{\text{chem}}\right)$. In the activated pathway, the hydrogen molecule acts as a precursor state ($H_{2,\text{physisorption}}$) and passes over the activation barrier state, accompanied by dissociation and the formation of a $M - H_{\text{chem}}$ bond.

4.4 HYDROGEN ADSORPTION IN POROUS MATERIALS

The adsorption of hydrogen can occur under both gas-phase (very low pressure) and electrochemical (aqueous or non-aqueous electrolyte providing a source of H) conditions. Very low temperatures are required to condense hydrogen (the critical point of hydrogen is about 33 K and 13 bar, with its boiling point near 20 K). Hydrogen adsorption in porous materials has the potential to dominate storage systems. These physical properties are related to weak $H_2 - H_2$ interactions. As hydrogen molecules contain no charge, no dipole moment, and a relatively weak quadrupole moment, and have low polarizability, the lack of significant interactions is inevitable. These same characteristics lead to the well-documented weak physical interaction of molecular hydrogen with solid surfaces. Nevertheless, it is possible to adsorb relatively large quantities of hydrogen at super-critical temperatures using materials with very high surface areas and very small pores within a narrow size range. The factors affecting the adsorption of hydrogen by porous materials include the strength of hydrogen–adsorbent interactions, the specific surface area (SSA), typical pore widths or diameters, and total pore volume. The strength of hydrogen–adsorbent interactions depends on the mutual properties of both hydrogen and the solid surface. The specific surface area defines the extent of the hydrogen–solid interface and the quantity of surface sites available for adsorption (per unit mass of adsorbent). It should be emphasized that small widths and sufficiently narrow pores significantly affect the adsorption capacity. Therefore, at low pressures, the amount of adsorption depends mainly on pore width or diameter rather than overall pore density. In very narrow pores, the overlapping potentials of opposing pore walls and neighboring framework atoms cause stronger solid–hydrogen interactions compared to open and flat surfaces. This allows for a larger amount of adsorption at relatively low pressure. The saturation limit of a solid substrate at high pressure tends to scale with the total pore volume up to pore sizes with no significant adsorption.

The hydrogen adsorption behavior is typically characterized by the shape of isotherms in relation to pressure and temperature. It is also affected by the physical properties of hydrogen (at operating T and P) and the intrinsic properties of the adsorbent. Hydrogen adsorption at sub-critical temperatures differs from super-critical adsorption. The SSA approximations are conducted at temperatures lower than the critical temperature of hydrogen to ensure the formation of a monolayer on relatively flat or open surfaces, followed by multilayer formation at higher relative pressures P/P_0, where P is absolute pressure and P_0 is the saturation pressure of the adsorbents. This approach is adopted specifically for non-porous or macroporous materials to identify the point at which a statistical monolayer is formed.

Solid materials as adsorptive hydrogen storage substrates are microporous or mesoporous (according to International Union of Pure and Applied Chemistry guidelines) [7]. Microporous hydrogen adsorbents consist of pore sizes <2 nm; mesoporous materials have pore sizes of 2–50 nm, and nanoporous materials contain pore sizes <100 nm. Therefore, porous hydrogen storage materials should be chosen from nanoporous candidates. In a typical mesoporous material at sub-critical temperatures, monolayer and multilayer adsorption occur, along with capillary condensation. This behavior creates characteristic (Type IV) isotherm shapes, with a sudden increase in the adsorbed amount and hysteresis loops between adsorption and desorption isotherms. In microporous hydrogen adsorption, the adsorption process is different, as adsorption only occurs via pore filling without capillary condensation, leading to a Type I hydrogen adsorption isotherm. The behavior is concave toward the pressure horizontal axis without observable hysteresis. This difference in adsorption behavior (capillary condensation versus pore filling) between Type IV and Type I is the basis for the arbitrary 2 nm threshold between micropores and mesopores. At super-critical temperatures, capillary condensation, a phase transition from a gas to a liquid-like state, cannot occur in mesoporous adsorbents. The presence of free surface area affects the density of the hydrogen adsorbate in the pores, which depends on pore size and the strength of the adsorbate–hydrogen interactions. The magnitude of the interaction between hydrogen and a surface or a pore of a certain size is usually characterized by the heat or enthalpy of adsorption (ΔH) or, in its most applicable form, the isosteric enthalpy of (ΔH_{iso}); estimated through Clausius–Clapeyron equations [8]. For instance, hydrogen on an open ceramic surface has ΔH_{iso} in the 3–4 kJ/mol range, which is low compared to the thermal energy of 1 kJ/mol. The ΔH_{iso} value increases with heterogeneous structures that alter the strength of interactions. Independent of the intrinsic chemistry of the adsorbent, there exists an increasing trend of enthalpy with decreasing pore size, up to the point of molecular sieving, which results from the enhanced interaction within small pores due to potential overlap. Electrostatic interactions are another operating force that can enhance hydrogen adsorption. This interaction increases ΔH_{iso} of hydrogen adsorption to 10 kJ/mol in open metal sites, MOFs, and cations in zeolites. It should be noted that in hydrogen adsorption, ΔH_{iso} tends to decrease as a function of coverage (θ), which significantly differs from systems that exhibit strong adsorbate–adsorbate interactions at higher coverages.

In hydrogen adsorption, the highest values of ΔH_{iso} are achieved in strongly interacting sites or the narrowest pores as indicated by low coverage systems. The ΔH_{iso} value at zero coverage can be calculated by extrapolation or direct calculation. This is, in principle, the energy release due to single molecule adsorption $(\theta \approx 0)$. The ΔH_{iso} value has practical applications such as (i) hydrogen adsorption can be carried out at higher temperatures for higher values of ΔH_{iso}; (ii) more heat can be generated upon adsorption when ΔH_{iso} is increased, which potentially raises heat management necessities; and (iii) the loading dependence of ΔH_{iso} indicates the deliverable amount of hydrogen at any pressure from a hydrogen storage perspective. There are high values of ΔH_{iso} at low loading contents which regardless of the value at high loading, can lead to more hydrogen being trapped in the substrate at low pressures (below the delivery pressure of the storage material).

Other ΔH_{iso} affecting parameters include the heterogeneity in the adsorbent due to functional groups, heteroatoms, different surface sites, surface defects, varying pore sizes within the adsorbent, and isotherm shape. The most applicable model is based on the Langmuir model, in which each adsorption site is equivalent (based on energy) without hydrogen–hydrogen interactions. Here, ΔH_{iso} as a function of loading is constant. Most real hydrogen adsorbents are non-homogeneous and cannot be adequately described by the Langmuir equation. The experimental hydrogen adsorption data can rarely be fitted for nanoporous substrates to an acceptable extent except over a limited pressure range.

As discussed, physical adsorption is generally based on van der Waals' interactions as *intermolecular interactions*. It includes long-range attractive forces between fluctuating molecules, permanent electric moments, induced electric moments, and short-range strong repulsive forces from the interaction of overlapping atomic or molecular orbitals (Pauli's exclusion principle) [9]. London dispersion forces are related to electron density fluctuations within atoms, which induce electrical moments in adjacent atoms, thus creating attraction. The absolute potential energy of attractive interactions between atoms can be estimated as follows:

$$|\varepsilon_d(r)| = A_1 r^{-6} + A_2 r^{-8} + A_3 r^{-10} \tag{4.39}$$

for r as the distance between the centers of the atoms, A_n are the dispersion constants for immediate dipole–dipole, dipole–quadrupole, and quadrupole–quadrupole interactions (n: 1, 2, 3) where the terms smaller than r^{-6} are negligible. The short-range repulsion (positive) potential energy can be calculated as follows:

$$\varepsilon_R(r) = B r^{-m} \tag{4.40}$$

where B is an empirical constant and m is about 12. Therefore, the total potential energy between two hydrogen atoms as a function of their distance can then be approximated using the Lennard–Jones (LJ) potential equation:

$$\varepsilon_{LJ}(r) = \varepsilon_R(r) + \varepsilon_R(r) = -Ar^{-6} + Br^{-12} \tag{4.41}$$

which has the general shape. At the van der Waals-type diameter σ where $\varepsilon_{LJ}(\sigma) = 0$, and interatomic distance (r_0) with the strongest interaction:

$$\varepsilon_0 \text{ at min}$$

$$\left(\frac{\delta\varepsilon_{LJ}(r)}{dr}\right)_{r_0} = 0$$

$$\varepsilon_{LJ}(r) = -4\varepsilon_0\left(\left(\frac{\sigma}{r}\right)^{-6} + \left(\frac{\sigma}{r}\right)^{-12}\right) \tag{4.42}$$

for

$$\varepsilon_0 = \varepsilon_{LJ}(r_0) = -\left(\frac{A}{4}\right)\sigma^{-6} \tag{4.43}$$

and

$$\sigma = -\sqrt[6]{\frac{B}{A}} \quad (r_0) = \sqrt[6]{\frac{2B}{A}} = \sqrt[6]{2\sigma} \tag{4.44}$$

For two atoms of i and j, the Lorentz–Berthelot mixing rules should be used [10]:

$$\sigma_{ij} = 1/2(\sigma_i + \sigma_j)$$

$$\varepsilon_{ij} = \sqrt{\varepsilon_i\varepsilon_j} \tag{4.45}$$

As the dispersion constant for atoms is directly related to their properties, it can be estimated through the Kirkwood–Muller approximation [11]:

$$A_{ij} = \frac{6mc^2\alpha_i\alpha_j}{(\alpha_i/\chi_i) + (\alpha_j/\chi_j)} \tag{4.46}$$

where α is the polarizability, χ is the magnetic susceptibility, m is the electron mass, and c is the speed of light. As an example, the α and χ values for hydrogen are 0.79×10^{-24} and 4.0×10^{-4} cm^3, respectively.

The LJ expression is a general potential term and is not limited to atoms; it can be used for any pair of particles. For instance, the position-dependent energy landscape that a molecule i experiences when close to an open surface or inside a porous framework comprising j atoms can be represented as LJ potentials:

$$U_{LJ}(r) = -4\varepsilon_0 \sum_J \left(\sigma_{hj}/r_{ij}\right)^{-6} + \left(\sigma_{ij}/r_{ij}\right)^{-12} \tag{4.47}$$

In homogeneous flat surfaces, the hydrogen–solid interaction potential is similar to the typical pairwise interaction curve. The new variable, instead of interatomic distance, is the distance from the surface (z). As a result of hydrogen confinement in pores, such as slits and cylinders, the interactions of adjacent atoms of the solid produce a much deeper potential well compared to open surfaces. In simple pore adsorption systems (i.e., slit-shaped or cylindrical pores), the energy dependency on distance creates deeper potential wells. The confinement of a hydrogen molecule in a 3D porous solid is more complicated, as the energy profile varies in the x, y, and z directions. A schematic representation of pore types, distance dependency ε_{ij}, and the 3D porous solid substrate is shown in Figure 4.6.

In hydrogen adsorption systems, the interactions become more complex as the adsorption temperature decreases. Additionally, hydrogen is a very light gas and therefore has a considerable de Broglie wavelength. It should be noted that particles exhibit wave-like properties. The de Broglie wavelength of a particle indicates the length scale at which wave-like properties are significant for that particle. The de Broglie wavelength is represented by λ_{dB}. For a particle with momentum p, the de Broglie wavelength is defined as follows:

$$\lambda_{dB} = hp \tag{4.48}$$

where h is the Planck constant. The concept of a circular orbit where de Broglie wavelengths fit into the circumference of the orbit for a typical hydrogen atom is shown in Figure 4.7.

As a result, nuclear quantum effects are expected to contribute significantly to the adsorption process at low temperatures. For such processes, the quadratic term of the Feynman–Hibbs effective potential is used to explain the deep part of the potential curve as follows:

$$U^{ij} = U_{LJ} + \left(\frac{\beta h^2}{96\pi^2 \mu_m}\right) \nabla^2 (U_{LJ}) \tag{4.49}$$

where $\beta = (kT)^{-1}$ and μ_m is the reduced mass:

$$\mu_m = \left(\frac{1}{m_i} + \frac{1}{m_j}\right)^{-1} \tag{4.50}$$

where m_i is the mass of certain hydrogen atom. For the reduced de Broglie wavelength (λ^*) and m as the mass of hydrogen,

$$\lambda^* = \left(\frac{\beta h^2}{2\pi m \sigma_{H_2}^2}\right)^{\frac{1}{2}} \leq 0.5 \tag{4.51}$$

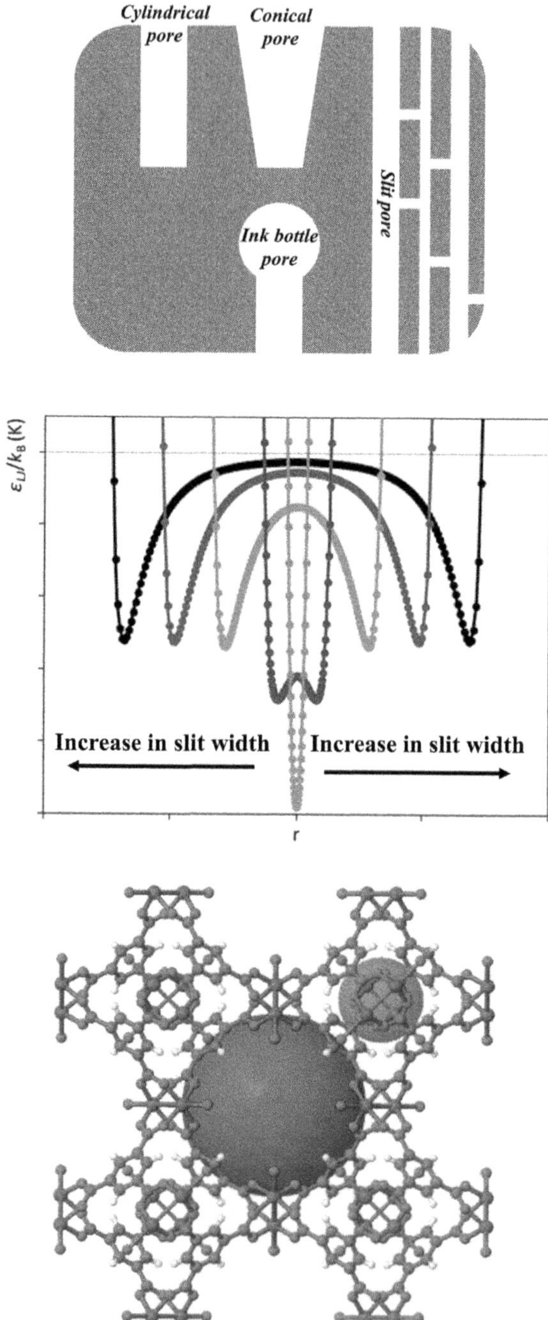

FIGURE 4.6 Typical confining pore structures for hydrogen adsorption, showing the position dependency of potential energy for a hydrogen molecule adsorbed in a typical slit pore with increasing width, along with the typical structure of a complicated 3D arrangement. Created, designed, and introduced by the author.

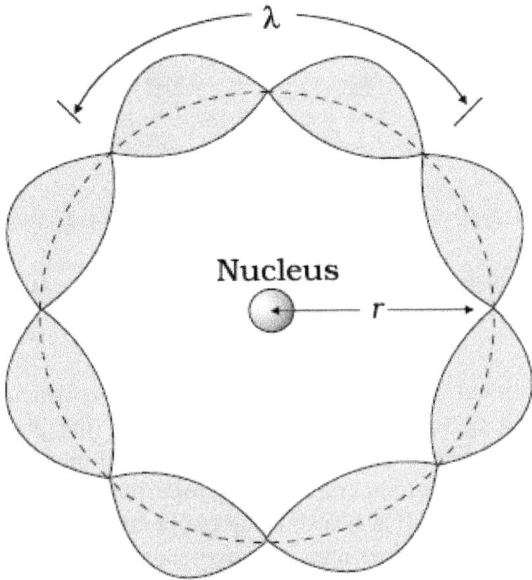

FIGURE 4.7   A wave on a circular orbit where four de Broglie wavelengths fit into the circumference of the orbit. Created, designed, and introduced by the author.

the above mentioned expression indicates quantum spreading precisely. The reduced thermal λ^* is estimated as $4.17 / \sqrt{T}$ (Feynman–Hibbs). In low temperatures, quantum effects are very important, and integral formalism explains the hydrogen interactions.

The low molecular mass of hydrogen and weak intermolecular potential result in significant properties at low temperatures for the adsorbed phase. Similar to other molecular/atomic gases near critical points, the adsorption of hydrogen at low temperatures is associated with the pore structure and morphology, involving processes like pore filling and capillary condensation, which are not probable at super-critical temperatures. In addition to this typical behavior, hydrogen adsorbed at low temperatures (around the boiling point) exhibits unique quantum behavior, which should be considered for the study of adsorption. For instance, the difference in kinetics and adsorption energy upon isotopic (hydrogen/deuterium) exchange depends significantly on the adsorption potential and pore size. The small difference in mass between these isotopes leads to quantum sieving in ultra-microporous (<0.7 nm) materials due to the larger de Broglie wavelength of the lighter isotope (deuterium transports faster than hydrogen in small pores) or enhanced chemical affinity for the heavier isotope at strong adsorption sites. This factor changes the kinetics of adsorption drastically. Moreover, the H_2 hydrogen molecule contains two possible proton spin state configurations: ortho-hydrogen (parallel, ↑↑) and para-hydrogen (anti-parallel, ↑↓). Due to the symmetrical wavefunction, para-hydrogen only has 2n rotational states ($J=0$, 2, 4,…), while ortho-hydrogen only has $2n+1$ states ($J=1$, 3, 5,…). The para state represents the lowest energy state, and ortho-hydrogen represents the highest energy state. Figure 4.8 shows the concept schematically.

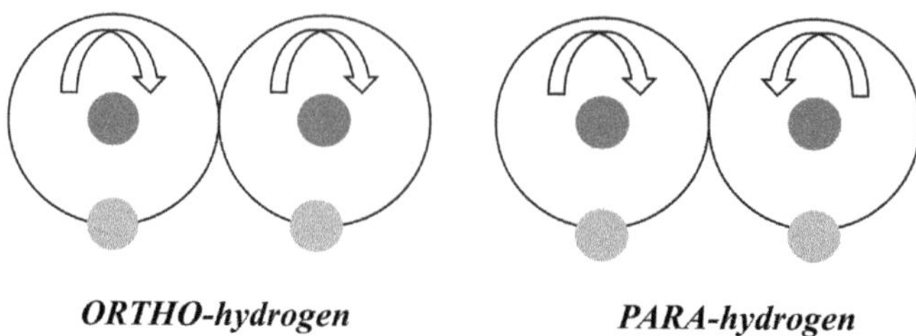

FIGURE 4.8 A schematic depiction of para and ortho-hydrogen states. Created, designed, and introduced by the author.

The lowest rotational state of ortho-hydrogen is $J = 1$, which indicates a quadrupole moment leading to a strong interaction during the physisorption process (molecules in the symmetric spin triplet state cannot fall into the lowest symmetric rotational state). At ambient temperature, the para/ortho ratio is estimated to be 1:3. This ratio changes to near 1:0 at near-boiling temperatures. If a hydrogen molecule is in the presence of a magnetic center, spin interchange may occur, which results in ortho–para conversion. The practical implications of this conversion are significant due to the large latent heat of conversion of 1.42 kJ/mol. This enthalpy is higher than the latent heat of evaporation of 0.89 kJ/mol at a similar temperature range. This effect can be neglected due to the complications and difficulties of controlling the para–ortho conversion and the low temperatures. The schematic temperature density of rotational energy for different spin states is presented in Figure 4.9. The quantum nature of molecular hydrogen is observed in the high compressibility of the bulk solid phase due to the absence of multiple electron shells.

Applying high pressure (typically about 100 bar) to hydrogen results in a 5% volume decrease, whereas other gases are less compressible (for example, argon shows changes in volume of less than 0.8%) (see Figure 4.10). Notice the difference between estimation and actual monolayer hydrogen adsorption.

This large compressibility also occurs in the adsorbed phase of hydrogen, which, due to short hydrogen–hydrogen intermolecular distances, can create densities higher than the bulk at sub-critical temperatures and near ambient pressure.

At *extra low temperatures*, hydrogen physisorption is primarily controlled by the solid surface area of the adsorbent, with a linear correlation observed between the SSA (area per unit of mass) and the hydrogen gravimetric capacity at elevated pressures, as explained by Chahine's rule. According to this rule, there is a 1 wt.% hydrogen adsorption for every $500\,\mathrm{m^2/g}$ of surface area. Therefore, the gravimetric and volumetric adsorption can be related to the surface area of the adsorbent with an average surface hydrogen density (constant). The estimated surface density is smaller than the bulk density of hydrogen and can be understood as the number of hydrogen molecules per unit area at a similar intermolecular (H_2–H_2) distance of 4.74 Å. This distance is larger than the intermolecular distances in solid (3.76 Å) and liquid (4.05 Å) states. The increase in the adsorbed layer density (or reduction of the intermolecular (H_2–H_2) distance) directly relates to an increase in the volumetric and gravimetric hydrogen storage capacity of a material. The monolayer capacity, combined with the cross-sectional area of the gas molecule

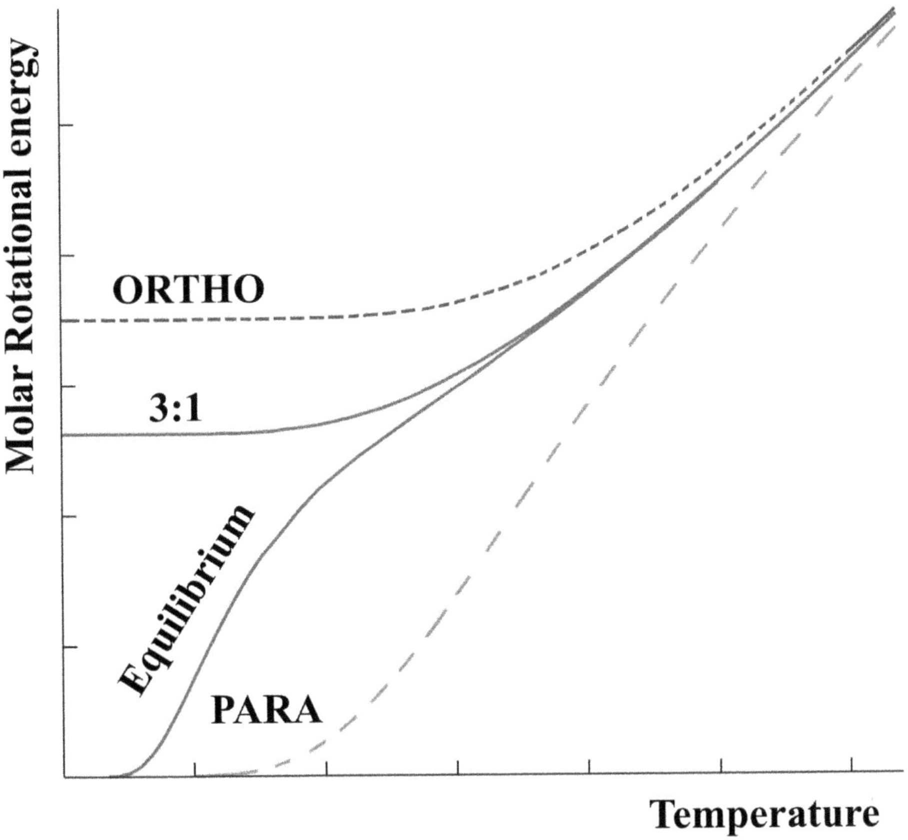

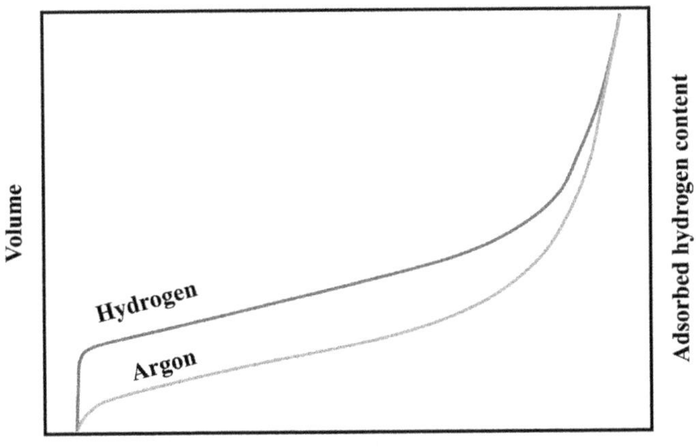

FIGURE 4.9 Molar rotational energy changes with temperature for spin states. Created, designed, and introduced by the author.

FIGURE 4.10 Comparison of typical monolayer adsorption isotherms between hydrogen and argon. Created, designed, and introduced by the author.

(the surface area a single molecule uses on the surface), can be related to the SSA. For instance, by knowing the cross-sectional area of the hydrogen molecule and applying sub-critical hydrogen adsorption principles, the maximum capacity of an adsorbent can be estimated. Hydrogen molecules exhibit very weak intermolecular interactions, with binding energies of ~0.3–0.5 kJ/mol, while the hydrogen–surface interaction potential for conventional materials such as oxide ceramics (carbon or silica) is about 4 kJ/mol. The density of the adsorbed layer or cross-sectional area at low temperatures depends strongly on the hydrogen molecule–surface interaction, which is essentially determined by both the surface chemistry and morphology of the adsorbent. The sub-critical adsorption of hydrogen is considered unusually higher than that of other gases in the first adsorbed layer (monolayer). The concept of double-layer adsorption (bilayer) can be used to explain high monolayer capacities on some adsorbents, such as carbon layers. According to theories, a short intermolecular H_2–H_2 distance (2.95 Å) is the main reason for bilayer formation, resulting in a hydrogen-adsorbed layer with approximately double the density. With high hydrogen adsorption capacity at sub-critical temperatures, factors like high density and monolayer-like hydrogen phase are considered mechanisms for bilayer formation. For example, the density of a single layer of hydrogen adsorbed on a series of micro-, meso-, and non-porous ceramics (silicas and carbons) demonstrates layer formation of hydrogen with an intermolecular distance of 2.9 Å. The isotherms indicate that hydrogen monolayer capacity is almost double that of other gases for a similar surface. The calculated intermolecular H_2–H_2 distance (based on hydrogen monolayer capacity and surface area) is about three times the bulk solid density of hydrogen (80 g/cm³) and around 200 g/cm³. The high density and short intermolecular H_2–H_2 distance are used to improve the storage capacities of materials for storage systems.

Based on this information about high capacity at low temperatures, storing hydrogen as a *liquid* in a cryogenic tank (at temperatures near 33 K and moderate pressures of about 13 bar) can be proposed as a viable option for large-scale storage for transport applications. Consequently, hydrogen adsorption in nanoporous materials such as MOFs (with bi-, trimodal, and even multimodal pore size distributions) near the boiling point could be used to increase the efficiency of cryogenic storage systems. At pressures close to the condensation point, the adsorbent material will become saturated due to the filling of all its pores. At this state, the loading reaches its upper limit, indicating the saturation capacity of the adsorbent material (cryo-adsorption). Saturation capacity is essential since it determines the upper physical limit of an adsorbent-filled tank in an isochoric (constant volume) adsorption, with direct measurement of the pressure increase with temperature. The immediate improvement from using a porous adsorbent includes a temperature increase from 20 to 40 K at the onset of the pressure increase point (wider temperature range).

During *adsorption at high pressures*, both the excess and absolute adsorbed quantities should be considered.

When measuring *hydrogen adsorption at high pressures*, it is important to distinguish between the excess and absolute adsorbed quantities. The excess amount is the adsorbed hydrogen over and above the molar quantity that would be present in the absence of gas–solid interactions. Meanwhile, absolute adsorption refers to the total hydrogen quantity present in the adsorbed phase. Furthermore, total adsorption is the sum of hydrogen adsorbate molecules in the accessible pore volume of the adsorbent. Theoretically, the volume occupied by the adsorbate or its density is essential for calculating absolute uptake from excess uptake. To determine the density of hydrogen adsorbed in the pores of an adsorbent, one continues to increase the pressure in an isotherm until adsorption saturates. As there is no further adsorption at higher pressure ranges and the hydrogen adsorbate density is constant, the calculated isotherm uptake will be linear with respect to the hydrogen bulk gas density, and the slope of the linear part of the isotherm indicates the volume of the adsorbate. Since hydrogen is typically adsorbed relatively weakly on most porous solid substrates, reaching saturation conditions requires low temperatures, high pressure, or a combination of the two. Because of the difficulties involved in handling and measuring high-pressure hydrogen adsorption, low temperatures are easier to apply for density investigations. By applying pressures below 100 bar at temperatures below 100 K, saturation can be achieved. At higher temperatures, even greater pressures are required; saturation cannot be reached at ambient temperatures for pressures in the 500–200 bar range in various nanoporous ceramics, including porous carbons, different MOFs, and hyper-crosslinked polymers (HCPs). Through hydrogen adsorption at critical temperatures, adsorbed hydrogen densities were estimated to be in the range of 0.05–0.07 g/cm^3. The typical behavior of hydrogen adsorbate in the excess isotherm, with calculated uptake versus gas pressure, absolute isotherm, and the density of hydrogen adsorbate calculated by absolute mass adsorbed divided by the adsorbate volume, is shown in Figure 4.11. Assuming that the adsorbed phase volume is constant for all points on the isotherm and that the hydrogen adsorbate density changes with adsorption, the absolute isotherm can be constructed from the excess isotherm by adding an uptake amount equivalent to the bulk gas density multiplied by the adsorbate volume. Similarly, while the adsorbate volume is kept constant, the density of the hydrogen adsorbate can be calculated from the absolute mass adsorbed

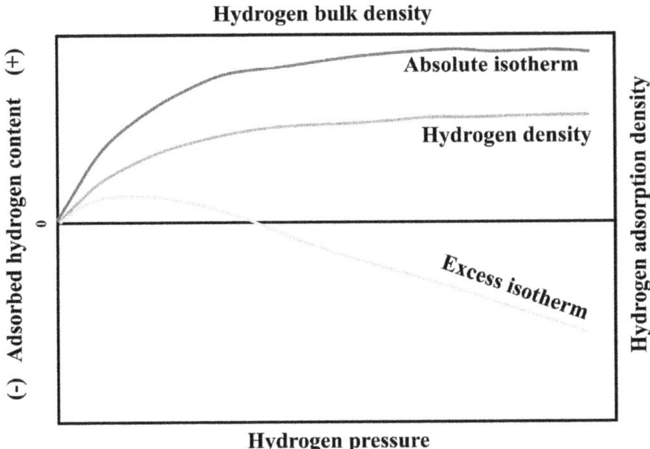

FIGURE 4.11 Adsorbed hydrogen content versus pressure for absolute isotherm, hydrogen density, and excess isotherm. Created, designed, and introduced by the author.

divided by the adsorbate volume. By fitting the excess uptake versus gas density isotherm with an acceptable regression coefficient (R), the maximum uptake and the maximum averaged density of adsorbed hydrogen can be estimated.

The theoretical and experimental data have revealed that combinations of low temperatures and moderate pressures guarantee hydrogen adsorption on porous adsorbents. Despite this limitation, various measures of the hydrogen storage capabilities of a material have been designed, and the main information for calculation is gathered from the uptake of hydrogen as a function of pressure. This information indicates the gravimetric capacity and, together with knowledge of the volume occupied by the sample, the volumetric capacities.

It is essential to analyze the source of hydrogen in storing environments in order to understand the unique mechanistic routes that operate. Some differences between electrochemical and gas-phase hydrogen include (i) that under gas-phase conditions, hydrogen is in the form of molecules and (ii) that under aqueous electrochemical conditions, the hydrated H^+ (acidic electrolytes) or the $H_2O(l)$ molecule (basic electrolytes) are the sources of electro-adsorbed hydrogen.

Electrochemical hydrogen absorption is considered reversible. The measurements of hydrogen uptake in the gas phase at low temperatures show similar quantities to the electrochemical capacities at ambient temperature. During the electrochemical charging process, hydrogen atoms accumulate at the surface of the electrode when electron transfer from the dispenser to the water molecules occurs. The hydrogen atoms can reassemble to form hydrogen molecules. This process continues until the adsorbent surface is completely covered with a monolayer of physisorbed hydrogen molecules [12]. Additional hydrogen molecules may not interact with the weak attractive forces of the surface. These hydrogen molecules can move freely and even form gas bubbles, which will be released from the electrode surface. The formation of a stable monolayer of hydrogen at the electrode surface at room temperature occurs if either the hydrogen atoms or the hydrogen molecules are motionless. In this condition, their surface diffusion must be kinetically mitigated by a large energy barrier, possibly due to the adsorbed electrolyte water molecules in the second layer. Another possible reaction path arises from the tendency of hydrogen atoms to chemisorb at the exterior surface of an adsorbent. The atoms move, flip, and recombine to form hydrogen molecules at high coverage, creating a concentric cylinder in the cavity of the adsorbent. If the binding energy of the chemisorbed hydrogen molecules is low compared to the energy in compounds, the absorbed amount of hydrogen remains proportional to the surface area of the adsorbent and could desorb at a relatively positive electrochemical potential. The adsorbed hydrogen content is proportional to the surface area of the nanoporous adsorbents at low temperatures. The geometric structure of the nanoporous adsorbents has no effect on the amount of absorbed hydrogen, as features such as curvature of porosities may only influence the adsorption energy and hardly affect the amount of adsorbed hydrogen. Besides the nanostructured adsorbents, other nanoporous materials can be used for hydrogen absorption. For instance, the hydrogen absorption of silicates with different pore architectures and compositions shows that hydrogen can be absorbed at the desired temperatures and pressures. Subsequently, hydrogen release upon heating these materials can be expected due to the absorption temperature. In these structures, the absorbed amount of hydrogen increases with increasing temperature and pressure. The adsorption behavior indicates that absorption in silicate structures is caused by a chemical reaction rather than physisorption. Due to their weak interactions, silicates cannot fulfill the volumetric storage targets, but the enthalpies of adsorption are significantly low; therefore, these materials can be used for hydrogen storage at ambient temperature and pressure. Aluminosilicate structures are hydrogen-trapping materials. In their structure, hydrogen can be trapped in the cavities of the molecular sieve at low temperatures and released by raising the temperature. At extremely low temperatures, these structures physisorb hydrogen in proportion to the specific surface area of the material. The low-temperature physisorption (Type I isotherm) of hydrogen in these structures is consistent with the adsorption model mentioned for nanostructured structures [13]. The desorption isotherm follows the same path as the adsorption, indicating that no pore condensation occurs. Microporous MOFs demonstrate that they absorb hydrogen at ambient temperature, which is directly proportional to the applied pressure. The micropores in aluminosilicates and their kinetic diameter control hydrogen storage performance. Additionally, the specific surface area and pore volume, the interaction of molecular hydrogen with the internal surfaces of the micropores, the stability of molecular movements, and the optimal storage temperature are the most important controlling parameters in the hydrogen storage properties of aluminosilicates. The hydrogen storage capacity of aluminosilicates depends on the base structures and their respective cations.

The natural clay is an example of porous material (such as montmorillonite and halloysite nanotube) with layered (tetrahedral–octahedral–tetrahedral) and rounded (tetrahedral–octahedral) structures. The modified clays can be used as hydrogen storage systems with multistage hydrogen sorption mechanisms and a linear increase in hydrogen storage content with increasing SSA. Their significant surface area often ranges from several hundred to over a thousand square meters per gram. The spillover mechanism (molecular hydrogen dissociation and the transfer of atomic hydrogen onto the adjacent support, potentially increasing storage capacity) is another feature of their adsorption. Another feature is the ionic exchange characteristic within the structure, which further improves their storage capacity by modifying the valence state and size of the exchangeable cations. This mechanism controls the diameter of the cages and channels and makes it possible to manipulate the accessible space (alkaline and earth alkaline ions exchange). The low hydrogen storage capacity at ambient temperatures and pressures, along with the high energy required for hydrogen desorption, has limited their utilization.

The slope of the linear relationship between the gravimetric hydrogen density and the hydrogen pressure is small. No saturation of hydrogen absorption is expected, which is unlikely for any kind of hydrogen absorption process. At low temperatures, a moderate amount of adsorbed hydrogen is expected at very low hydrogen pressures, with a slight, almost linear, increase as pressure increases. This behavior is not exactly characteristic of a Type I isotherm. The significant advantages of physisorption for hydrogen storage are the low operating pressure, the relatively low cost of the involved adsorbents, and the simple design of the hydrogen storage system. However, the relatively small amount of adsorbed hydrogen on some ceramic adsorbents, such as carbon-based materials, along with the low necessary temperatures, are significant drawbacks of hydrogen storage based on physisorption. Other special adsorption properties can be attributed to extra high-surface-area structures, such as SP^2 hybridized 2D atomic crystals, i.e., graphene, black phosphorus, molybdenum dichalcogenides, and tungsten dichalcogenides. There are practical deficiencies in the utilization of these structures as hosts in electrochemical energy storage devices. Some of these deficiencies include (i) they have much lower capacitance than the theoretical maximum, (ii) their macroporous nature limits volumetric energy density, and (iii) the low packing density of structure-based electrodes. Black phosphorus is a 2D material derived from a relatively rare allotrope of phosphorus that is composed of puckered honeycomb layers stacked by weak van der Waals' interactions. Molybdenum dichalcogenides, such as molybdenum disulfide (MoS_2) and molybdenum diselenide ($MoSe_2$), are 2D graphene-like transition metal dichalcogenides that resemble graphite crystals. MoS_2 and $MoSe_2$ are weakly bonded by van der Waals interactions, exhibiting indirect-to-direct bandgaps. Various structures of 2D adsorbents, including graphene sheets (with and without defects), graphene oxide(GO), and dichalcogenides, are shown in Figure 4.12.

Other similar structures include copper phthalocyanine (CuPc), MOF-based hybrid frameworks (MOFs, porous organic polymers, covalent organic frameworks (COFs)), and HCPs.

Copper phthalocyanine contains a planar framework with four iso-indole units, with nitrogen atoms in an azo position linking these iso-indoles. These wide linkages lead to longwave radiation absorption, probably associated with a $\pi \rightarrow \pi^*$ transition. From an electrochemical aspect, superior hydrogen storage discharge capacity and high efficiency (charge–discharge) for rectangular nanocuboids of beta copper phthalocyanine (β-CuPc) are achievable. In a hybrid framework with barium aluminate (BAO), they serve as hosts for electrochemical hydrogen storage. It is conceivable

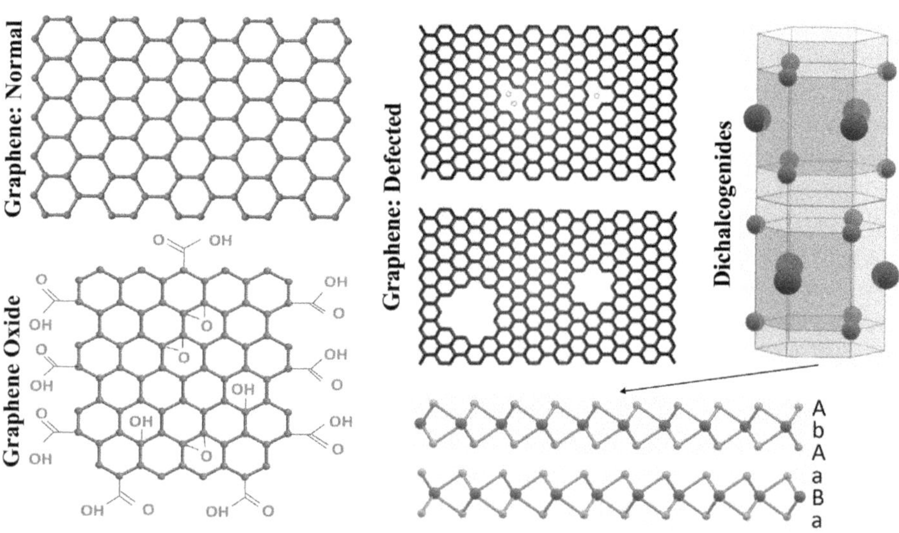

FIGURE 4.12 Various structures of 2D adsorbents and their stacking. Created, designed, and introduced by the author.

that the discharge capacity of BAO nanoparticles could be enhanced upon the addition of phthalocyanine by up to 50% improvement, as it is known that the presence of redox species like Cu(II) can enhance hydrogen storage [14].

Porous organic polymers, generally amorphous porous organic compounds, can be categorized as porous aromatic frameworks (PAFs), conjugated microporous polymers (CMPs), HCPs, and polymers of intrinsic microporosity (PIMs). PAFs exhibit elevated specific surface area, notable porosity, and strong π–π conjugation. Consequently, PAFs have applications in hydrogen gas storage [15]. Typical porous aromatics show high stability and exceptional surface area with Langmuir behavior. Most PAFs demonstrate significant hydrogen uptake capacities, especially when improved by substituting the central carbon atom of the tetrahedral monomers with adamantane ($C_{10}H_{16}$), silicon, and germanium as hydrogen storage media. These substitutions create surface areas of to $10,000\,m^2/g$ SSA and ~9wt.% adsorption capacity at low temperatures and moderate pressures [16].

CMPs are a class of organic porous polymers that combine π-conjugated skeletons with permanent nanopores, differing from other porous materials that are not π-conjugated as well as conventional conjugated polymers that are non-porous. The structure can provide more than $1200\,m^2/g$ SSA with significant ~10wt.% storage at very low temperatures [17]. The isosteric enthalpy for hydrogen adsorption is in the <10 kJ/mol range, which is well within the physisorption range. CMPs generally exhibit large surface areas and micropore sizes in the 0.5–1.5 nm range [18].

COFs are a class of hydrogen storage materials that are similar to MOFs but do not contain heavy metal ions. They are formed by organic building blocks through dynamic covalent formation methods. COFs consist of extended crystalline structures with high surface areas and low densities. Due to their low density, they provide an advantage in gravimetric hydrogen storage capacity. Depending on the dimensional arrangement of the hypothetical unit cells, the crystalline framework of COFs can be categorized as 2D or 3D (see Figure 4.13). These COF frameworks consist of extensive ordered structures and form a polycrystalline network due to the partially reversible nature of covalent bond formation during the condensation reaction (known as dynamic covalent chemistry). The hydrogen storage capacities of COFs (2D and 3D) under high pressure are significant, in the 7–8 wt.% range. Decorating COFs with metal atoms significantly improves their hydrogen adsorption capacity; for example, lithium decoration enhances the adsorption of three hydrogen molecules around lithium–magnesium atoms, resulting in nearly 7 wt.% hydrogen adsorption at room temperature, or four hydrogen molecules with scandium [19].

HCPs (also known as porous organic polymers) can also reversibly absorb and release hydrogen via physisorption. These structures have covalently bonded (hydro) and thermally stable backbones that exhibit high and accessible free surface areas. HCPs are essentially highly interconnected porous polymers. They can be synthesized through extensive post-crosslinking of linear polystyrene chains. The process includes the formation of structural bridges between adjacent aromatic rings and other chains in a highly swollen state. The resulting porous structures offer numerous advantages over non-porous counterparts due to their high porosity. As a result, hydrogen can be fixed and stored in N-heterocycle polymers at atmospheric pressure through the formation of chemical bonds, producing the corresponding alcohol

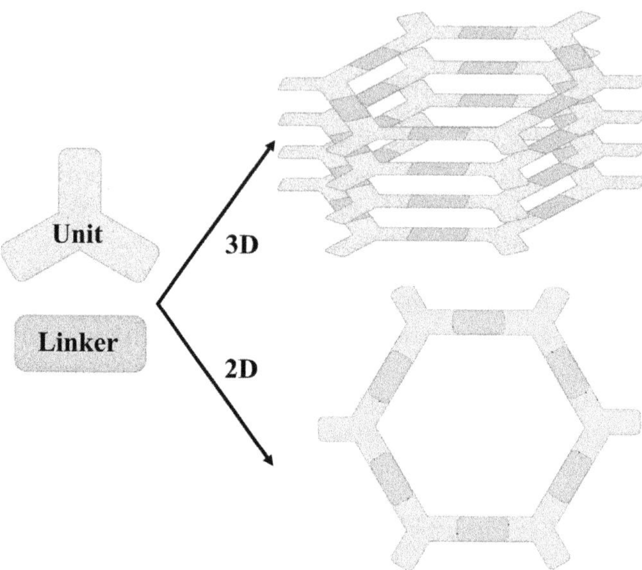

FIGURE 4.13   Typical depiction of COFs. Created, designed, and introduced by the author.

and hydrogenated N-heterocycle polymers. Electrochemical hydrogen sorption using water as a hydrogen source can also be effective, where the polymer acts as a scaffold for hydrogenation [20]. The hydrogenated polymers release hydrogen in the presence of catalysts under mild conditions. The potential of using organic polymers in the quest for new types of hydrogen-carrying and -storing materials that are safe and portable is promising. The concept of hyper-crosslinking in polymers and hydrogenation is shown in Figure 4.14. A typical HCP has a surface area of 1500–2000 m^2/g with a capacity of 3–5 wt.%.

As polymers can be crystalline, semi-crystalline, or amorphous, the hydrogen sorption physical properties can be related to the extent of crystallinity of the polymer. Highly crystalline polymers (<100% crystallinity) are very rigid and lack sorption sites (barrier effects) as well as the mobility of the chains to allow the mass transfer of gas molecules. Semi-crystalline polymers (50–75% crystallinity), on the other hand, contain more than one phase of crystalline and amorphous structures, making them less densely packed and providing more gas sorption sites. The amorphous polymers (or amorphous phase of semi-crystalline polymers) possess multiple micro/nanopores. These polymers have advantages over other porous structures, including high porosity, being metal-free, and being lightweight. HCPs, such as polystyrene, are macromolecules that can be fabricated via alkylation reactions. These HCPs are known to store hydrogen up to 5 wt.% at high pressures and low temperatures.

Fused-ring polymers (PIMs composed wholly of fused-ring subunits designed to provide highly rigid and contorted macromolecular structures that pack space inefficiently) are one class of organic materials with high surface areas (600–1800 m^2/g), micropores smaller than 2 nm, and very efficient hydrogen storage capabilities. They are characterized by rigid ladder-like backbones interspersed with sites of contortion,

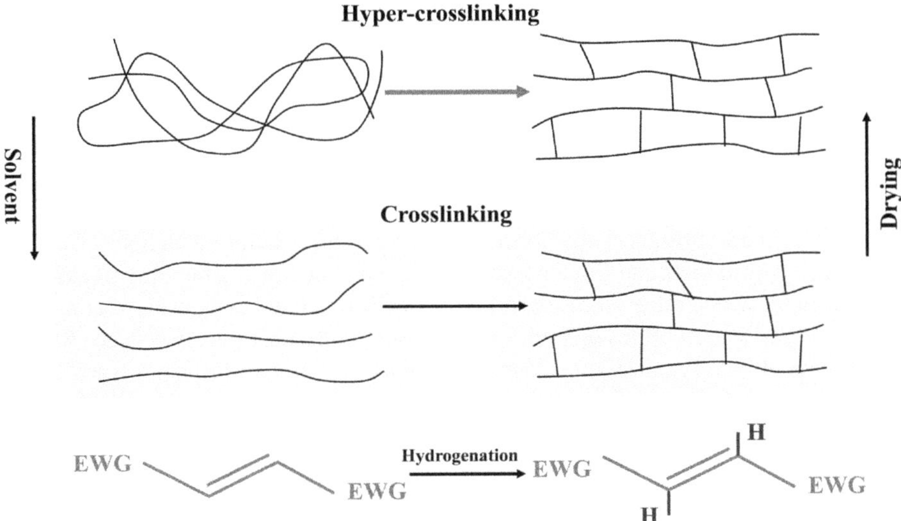

FIGURE 4.14 Schematic representation of the hyper-crosslinking and hydrogenation process. Created, designed, and introduced by the author.

which hinder efficient packing in the solid state, resulting in substantial free volume. There is a small but statistically significant decrease in the hydrogen storage capacity of a typical PIM in the initial stages of aging, which is attributed to the slow rearrangement of the polymer scaffolds. These polymeric networks can reversibly adsorb hydrogen up to 1.5 wt.% at low temperatures (this limitation is attributed to constrained surface area and pore volume) [18]. Similar to other adsorbents, fused-ring polymers can adsorb and desorb hydrogen under ambient conditions. By increasing the interaction energy of hydrogen with the polymer surface through reduced pore size and enhanced adsorption enthalpy, higher adsorption capacity can be achieved, allowing for reversible hydrogen storage at room temperature. Some of the polymer hydrogen adsorbents include polyaniline (PANI) and polypyrrole (PPy), alcohol-based polymers, cyclotricatechylene-based porous crystalline materials, hexachlorohexaazatrinaphthylene (HCLHATN), and tetrafluoroterephthalonitrile ($C_8F_4N_2$). The combination of *organic–inorganic* systems with (i) non-covalent interactions and (ii) some covalent interactions comprises another category of hydrogen storage systems. The materials involved in a large group of organic–inorganic materials are classified as COFs, microporous metal coordination materials, organo-transition metal complexes, and polymeric-based composites. In the composite/hybrid hydrogen storage materials, two or more components combine to overcome their weaknesses, and the hydrogen capacities in the composite materials reach values between those of the base components.

A *hydrogel of swelled polymers* (three-dimensional crosslinked hydrophilic polymers that can swell in an aqueous environment without dissolving, with a water content of 25–35 wt.%) can be utilized due to their potential in hydrogenation/dehydrogenation processes. High concentrations (1 M) of fluorenol/fluorenone units in the polymer or amorphous gel are used for these applications. The mechanism includes a strong reaction to hydrogen and efficient hydrogen-exchanging reactions, wherein units are fully hydrogenated and rapidly dehydrogenated throughout the polymer. Fluorenone $\left((C_6H_4)_2\,CO\right)$ and carbon fiber composite nanohybrids are another example of polymers used for hydrogen fixation/release in aqueous mediums. N-heterocyclic polymers such as quinaldine $(CH_3C_9H_6N)$ have potential as hydrogen carriers with approximately 3 wt.% hydrogen capacities. Quinaldine and 1,2,3,4-tetrahydroquinaldine composites are other examples of polymeric hydrogen storage system.

Clathrates are substances that form a specific crystal structure with holes of appropriate size within the three-dimensional structure, formed by the bonding of atoms or molecules, in which other atoms or molecules exist at a fixed composition ratio. They assume cage-like structures and encapsulate guest molecules (i.e., hydrogen) without any chemical bonding. Clathrates can be categorized into two principal types: clathrate hydrates and semi-clathrate hydrates, based on the interactions between the guest species and the host framework. Semi-clathrate hydrates exhibit a combination of partial hydrogen bonding and van der Waals force interactions between guest and host atoms, while clathrate hydrates primarily rely on weak van der Waals forces to govern the interaction between host and guest molecules [21]. The containment of hydrogen within cages, as facilitated by clathrates, helps to substantially reduce the associated hazards related to sudden gas release. Additionally, the utilization of minimal promoters in conjunction with water enhances operational conditions; the ability to store hydrogen in its molecular form allows for immediate utilization with minimal thermal or depressurization requirements; and the capacity to store hydrogen at moderate temperatures and pressures in cases involving low promoter content, as well as the promise of substantial gravimetric and volumetric storage capacities, are the main benefits of clathrates as adsorbents. Table 4.2 summarizes the details on structures and promoters of different types. The difficulty in forming clathrates under typical ambient conditions is related to hydrogen's small molecular size. The use of promoter molecules, such as tetrahydrofuran, has shown promising outcomes in enhancing hydrate structures for hydrogen storage. The presence of these promoter molecules facilitates the stabilization of the hydrate structure and significantly augments hydrogen storage capacity.

TABLE 4.2

Structures and Promoters of Clathrates Hydrogen Adsorbents

Clathrates	Structure	Cage type	Forming gas
Type I	Cubic	Small dodecahedral	Hydrogen
		Large tetrakaidecahedral	Methane
			Ethylene
Type II	Cubic	Small dodecahedral	Hydrogen
		Large hexakaidecahedral	Ethane
			Propane
			Carbon dioxide
Type III	Hexagonal	Small dodecahedral	Hydrogen
		Medium hexakaidecahedral	Kryptone
		Large icosakaioctahedral	Xenone

REFERENCES

1. Kumar A, Muthukumar P, Sharma P, Kumar EA. Absorption based solid state hydrogen storage system: A review. *Sustainable Energy Technologies and Assessments*, 2022, 52: 102204.
2. Chandra Muduli R, Kale P. Silicon nanostructures for solid-state hydrogen storage: A review. *International Journal of Hydrogen Energy*, 2023, 48(4): 1401–39.
3. Lu Q, Zhang B, Zhang L, Zhu Y, Gong W. Monolayer AsC5 as the promising hydrogen storage material for clean energy applications. *Nanomaterials*, 2023, 13(9): 1553.
4. Brakat A and Zhu H. From forces to assemblies: van der waals forces-driven assemblies in anisotropic Quasi-2D graphene and Quasi-1D nanocellulose heterointerfaces towards Quasi-3D nanoarchitecture. *Nanomaterials*, 2023, 13(17): 2399.
5. Sdanghi G, Schaefer S, Maranzana G, Celzard A, Fierro V. Application of the modified Dubinin-Astakhov equation for a better understanding of high-pressure hydrogen adsorption on activated carbons. *International Journal of Hydrogen Energy*, 2020, 45(48): 25912–26.
6. Yoshida K, Kajiwara K, Sugime H, Noda S, Hanada N. Numerical simulation of heat supply and hydrogen desorption by hydrogen flow to porous MgH2 sheet. *Chemical Engineering Journal*, 2021, 421: 129648.
7. McCusker LB, Liebau F, Engelhardt G. Nomenclature of structural and compositional characteristics of ordered microporous and mesoporous materials with inorganic hosts: (IUPAC recommendations 2001). *Microporous and Mesoporous Materials*, 2003, 58(1): 3–13.
8. Knight EW, Gillespie AK, Prosniewski MJ, Stalla D, Dohnke E, Rash TA et al. Determination of the enthalpy of adsorption of hydrogen in activated carbon at room temperature. *International Journal of Hydrogen Energy*, 2020, 45(31): 15541–52.
9. Yin Y-H, Wang Q. The electric fields enhance the non-covalent intermolecular interaction between H_2 and $(MgO)_3$. *Chemical Physics*, 2020, 539: 110925.
10. Shekhar S, Chowdhury C. Prediction of hydrogen storage in metal-organic frameworks: A neural network based approach. *Results in Surfaces and Interfaces*, 2024, 14: 100166.
11. Zhang L, Allendorf MD, Balderas-Xicohténcatl R, Broom DP, Fanourgakis GS, Froudakis GE et al. Fundamentals of hydrogen storage in nanoporous materials. *Progress in Energy*, 2022, 4(4): 042013.
12. Jerkiewicz G. Electrochemical hydrogen adsorption and absorption, Part 1: Under-potential deposition of hydrogen. *Electrocatalysis*, 2010, 1(4): 179–99.
13. Payzullakhanov MS, Parpiev OR, Avezova NR, Shermatov Z. Hydrogen storage in porous ceramic materials of aluminosilicate composition. *Applied Solar Energy*, 2022, 58(5): 722–4.
14. Salehabadi A, Morad N, Ahmad MI. A study on electrochemical hydrogen storage performance of β-copper phthalocyanine rectangular nanocuboids. *Renewable Energy*, 2020, 146: 497–503.
15. Cousins K, Zhang R. Highly porous organic polymers for hydrogen fuel storage. *Polymers*, 2019, 11(4): 690.
16. Tan L, Tan B. Hypercrosslinked porous polymer materials: design, synthesis, applications. *Chemical Society Reviews*, 2017, 46(11): 3322–56.
17. Hayat A, Sohail M, El Jery A, Al-Zaydi KM, Raza S, Ali H et al. Recent advances in ground-breaking conjugated microporous polymers-based materials, their synthesis, modification and potential applications. *Materials Today*, 2023, 64: 180–208.
18. McKeown NB, Gahnem B, Msayib KJ, Budd PM, Tattershall CE, Mahmood K et al. Towards polymer-based hydrogen storage materials: Engineering ultramicroporous cavities within polymers of intrinsic microporosity. *Angewandte Chemie International Edition*, 2006, 45(11): 1804–7.

19. Tan KT, Ghosh S, Wang Z, Wen F, Rodríguez-San-Miguel D, Feng J et al. Covalent organic frameworks. *Nature Reviews Methods Primers*, 2023, 3(1): 1.
20. Kato R, Nishide H. Polymers for carrying and storing hydrogen. *Polymer Journal*, 2018, 50(1): 77–82.
21. Gupta A, Baron GV, Perreault P, Lenaerts S, Ciocarlan R-G., Cool P et al. Hydrogen clathrates: Next generation hydrogen storage materials. *Energy Storage Materials*, 2021, 41: 69–107.

5 Hydrogen Storage
Ceramic Hydrogen Batteries

Scope

In this chapter, a brief description of the ceramic-based hydrogen storage system is provided. The systems are primarily based on the weak interaction of hydrogen molecules with the adsorbent surface within pores (sites) or active zones developed as porous materials for hydrogen sorption. The systems demonstrate how the structure and morphology play a crucial role in the hydrogen storage performance of the host adsorbent ceramic materials. The presented information is based on the amount of hydrogen adsorbed as a function of pressure, temperature, the enthalpies of adsorption, gravimetric/volumetric capacities, and the adsorption/desorption characteristics, with specific characteristics of non-metallic non-polymeric systems.

5.1 CERAMIC HYDROGEN ADSORBENTS

Through the concept of adsorption via ceramic materials, a combination of monolithic and composite ceramics with high surface area, with or without hydride compounds, acts as hydrogen carriers. A variety of systems have been developed to address the primary requirement of hydrogen storage capacity, and then consider the temperature and pressure profile of operation, reversibility and regenerability, non-toxicity, and availability. Table 5.1 summarizes representatives of these systems.

5.1.1 ALUMINOSILICATES

The adsorption ability of aluminosilicates, such as clay minerals and zeolites, is derived from their unique surface properties that allow them to absorb/adsorb hydrogen molecules, accompanied by a partial or complete swelling of their layered structure, followed by the sliding of layers relative to each other. For instance, microporous zeolites can achieve a storage capacity of ~2.1 wt.% at low temperatures and a pressure range of up to 1.6 MPa. The phenomenon of adsorption by aluminosilicates involves sorption, capture, or uptake, which can occur through direct contact with hydrogen. The adsorption properties of aluminosilicates can be determined by the state and chemical activity of their surface, porosity, and external conditions. Adsorption capacities and the physical–chemical properties of the surface of natural or treated aluminosilicates form the basis for natural state and reagent-treated (acid and alkaline activation) adsorbents. It should be noted that natural aluminosilicates are always a combination of different phases, such as clay and non-clay minerals, mixed phases, oxides, organic matter, and similar substances, which can form under varied geological conditions. For specific applications such as adsorption, separation,

DOI: 10.1201/9781003488828-5

TABLE 5.1

Ceramic Hydrogen Systems [1]

	Hydrogen storage capacity (wt.%)	Temperature (°C)/ pressure (bar)	Advantage	Limitations
Carbon allotropes	5–10	−196 to 25/ 60–100	High surface area	Low operation temperature
Aluminosilicates	2.5–4	−196 to −40/40	High surface area	Low operation temperature
Oxides	2–6	−196 to 120/ 20–200	High surface area	Medium capacity
MXene hosts	2–13	−196 to 25/ 25–60	High gravimetric capacity	Low operation temperature
Impregnated MXenes	5–14	25–500/ 1–30	High capacity	Low regenerability

and purification, it is necessary to deploy natural poly-mineral phases to enrich the content of a certain aluminosilicate phase or to eliminate admixtures. Some of these processing steps include acid modification, alkaline modification, and ion exchange.

The extent of acid/alkaline modification of aluminosilicates' chemical composition and surface properties depends directly on (i) the structural features of the major aluminosilicate phase in the material and (ii) the type of modifying agent and conditions. Octahedral sheets of 1:1 layered type phases are more accessible for acid modifications compared to 2:1 planar (montmorillonite and illite) or 2:1 non-planar (palygorskite) mineral phases. Tetrahedral sheets of all aluminosilicate types undergo chemical interactions with alkaline solutions, producing either an amorphous or crystalline aluminosilicate phase. The variations in chemical composition, as well as in the structure of modified aluminosilicates, affect the activity of the surface. Moreover, the nature of surface-active sites (adsorption and surface activity) of natural aluminosilicate samples, as well as the effects of surface modification by acid and alkaline solutions on their adsorption strength, should be considered.

Aluminosilicates are categorized as a group of hydrous aluminum, magnesium, and iron silicates that have layered or semi-layered structures. The main structural elements of most natural aluminosilicates are tetrahedra of SiO_4 (T) and/or octahedra of AlO_6 (O) attached at the vertices, forming planar sheets known as tetrahedral and octahedral sheets, respectively. Examples of hydrogen storage systems include planar phyllosilicates, which consist of T- and O-sheets formed in the elementary layer of the 1:1 type (kaolinite, dickite, etc.) and the extensive 2:1 type (talc, montmorillonite, vermiculite) where the octahedral sheet is sandwiched between two T-sheets (T-O-T layer). The building blocks and primary layers of typical aluminosilicates are represented in Figure 5.1.

The type of interlayer material, exchangeable cations of alkali/alkaline earth metals, hydroxides, and the layer charge ($x = 0$–2) allow for further subdivision of aluminosilicates into subgroups. Additionally, a group of non-planar minerals known as "quasi-layered silicates" exhibit a modulated structure (strips, islands,

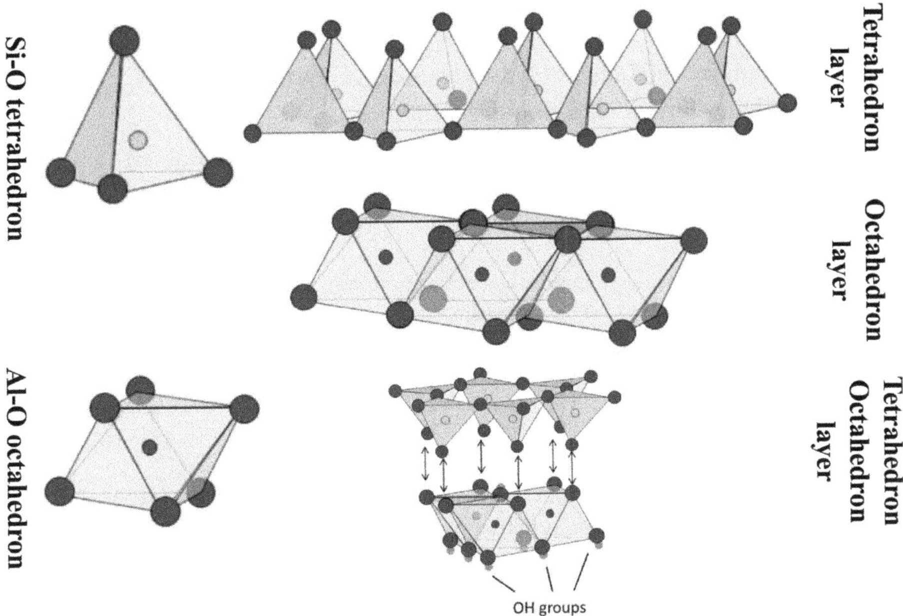

FIGURE 5.1 Si-O tetrahedron, Al-O octahedron, and primary layered building blocks of typical aluminosilicates. Created, designed, and introduced by the author.

ribbons). Unlike the planar types, discontinuous chains of T-sheets can be inverted, linking adjacent 1:1 or 2:1 sheets and producing zeolites with channel-like structures. The most remarkable cation-exchange capacity (CEC) is found in minerals of the 2:1 group, specifically layered-chain minerals (sepiolite, palygorskite). The kaolinite group possesses the lowest CEC, as well as low swelling due to weak isomorphic substitutions. The structure of clay minerals and their chemical composition can be expressed by $M_{2/n}OAl_2O_3ySiO_2zH_2O$ indicates active sites, which are responsible for the adsorption of aluminosilicates like zeolites. Processing includes exchangeable cations (K, Na, Li, Ca), hydroxyls of acid/base (SiOH, SiO(H)Al, Al-OH, Mg-OH), unsaturated ions of Al, Mg, Fe, and oxygen anions. Such reactivity of the surface and structural elements of silicates indicates their ability for physical adsorption through van der Waals' (vdW) interactions or hydrogen bonding with surface groups like unsaturated cations and oxygen anions. Furthermore, there is a proven opportunity for chemisorption due to strong interactions between the surface and the molecules, followed by the formation of chemical bonds. The ion-exchange reactions (involving exchangeable cations or OH groups) contribute to the adsorption process. The general properties of aluminosilicates and the possible required processing for enhancing hydrogen adsorption are listed in Table 5.2. As presented, the porosity, surface area, and surface physical–chemical affinity of aluminosilicates require improvement to meet the general requirements for hydrogen adsorption.

Modifications are carried out via chemical modification, ion exchange modification, and activations. Chemical modification of aluminosilicates has a direct

TABLE 5.2

Aluminosilicates as Hydrogen Storage Systems: Characteristic and Requirements

Property	Characteristic	Required Processing
Specific surface area	Low to medium	Site activation process
Porosity	Mesoporous	Chemical modification
		Thermal activation
Chemical affinity	Low for non-polar adsorbates	Ion modification
Chemical stability	Low to medium	Chemical modification
Cyclability	Acceptable	Ion modification
		Thermal activation

influence on surface properties. Aluminosilicates can easily interact with chemical reagents and be modified through their surface and structural properties, resulting in changes to their physical–chemical activity. This characteristic has enabled them to be tailored for targeted applications, including hydrogen adsorption. The activation and modification of aluminosilicate properties are carried out via activation methods that include mechanical (grinding via planetary ball milling, vertical jet, cryogenic milling), thermal, chemical (acid, alkali, salt reaction), and physical (exposure to magnetic/electric fields), as well as their combination into mechanochemical and thermochemical activation. Mechanical and mechanochemical activation results in the reduction of particle size (improving specific surface area (SSA) and varying the porosity content), deformation of the crystal structure, mainly along the c-axis, and dissociation of Si-OH, Al-OH, Al-O-Si, and Si-O bonds (active adsorption sites), potentially leading to amorphization of the structure. Mechanical modification is accompanied by the formation of new active sites at the boundaries of particle faults and an increase in CEC. Other significant changes to structure and properties can be achieved through mechanochemical treatment in chemical media (e.g., ionized water, acid solutions, and alkaline solutions). Thermal activation is based on thermal treatment at elevated temperatures, resulting in the desorption of physisorbed and chemisorbed gases and water molecules, which leads to the evolution of free surface.

Chemical activation modification directly affects the structure and properties due to chemical interactions between the modifying agent and the surface. Despite numerous modifying agents being introduced to alter the surface chemical profile of aluminosilicates, activation by acid and alkaline solutions remains the most important procedure. For instance, in acid activation or acid leaching, inorganic acids (such as sulfuric or hydrochloric acid) and organic acids (e.g., acetic, formic, and oxalic acids) are used. The chemical reactions that occur when aluminosilicates come into contact with an acid solution include ion-exchange reactions and acid attack. Furthermore, acid activation can modify aluminosilicates through the dissolution of cationic components, leading to an increased porous structure and SSA, and thereby improving adsorption capacity. The exchangeable cations include alkali or alkaline earth cations such as Na, K, and Ca, which are located in the interlayer space and can be replaced by protons (H$^+$) resulting from the dissociation of an acid to form hydrogen from aluminosilicate according to the following equation:

$$M_x^+((AlO_2)_x(SiO_2)_{n-x})^{x-} + x\left(H_2O\text{-}H^+\right) \leftrightarrow xM^+ + (H_3O)_x^+(AlO_2)_x(SiO_2)_{n-x})^{x-}$$

with Na, Mg, Ca, and K as M. Besides ion-exchange reactions, acid attack on the layered structure occurs as protons from the acid easily penetrate into the interlayer space and react primarily with the cations of the octahedral sheet (de-alumination), resulting in the substitution of octahedral aluminum or iron with protons and the formation of Si-OH groups in the tetrahedral sheet.

$$((AlOOH)_x^0(SiO_2)_{n-x})^0 + 3xH^+ \leftrightarrow ((H_4O_2)_x(SiO_2)_{n-x})^0 + xAl^{3+}$$

During short-term acid modification, ion-exchange reactions occur with slight destruction of the layered structure, which increases the pore size of aluminosilicates, transforms micropores into mesopores, and raises the SSA. In contrast, long-term acid modification destroys the crystal structure, leading to a decrease in the surface area and the adsorption capacity of aluminosilicates. Figure 5.2 schematically compares the changes in adsorption isotherms with loops of kaolinite and palygorskite, showing poly-molecular adsorption and capillary condensation before and after acid modification, with a significant increase in SSA (mesoporous with 2–25 nm pores).

The significant increase in the density of small-sized pores after acid modification is evident for most aluminosilicates (see Figure 5.3). The significance of acid modification on aluminosilicates involves an increase in pore density by two or three orders of magnitude, achieved through the additional formation and opening of mesopores

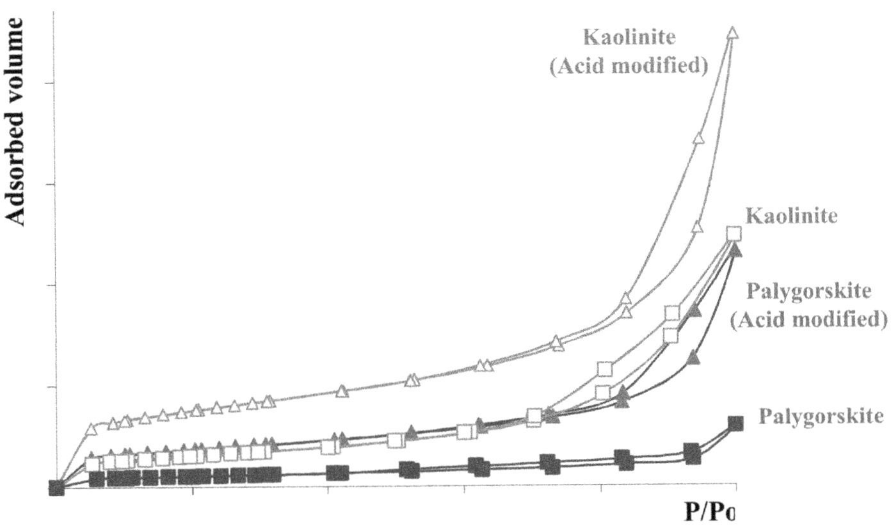

FIGURE 5.2 Schematic comparison of the effect of acid modification on aluminosilicates. Created, designed, and introduced by the author.

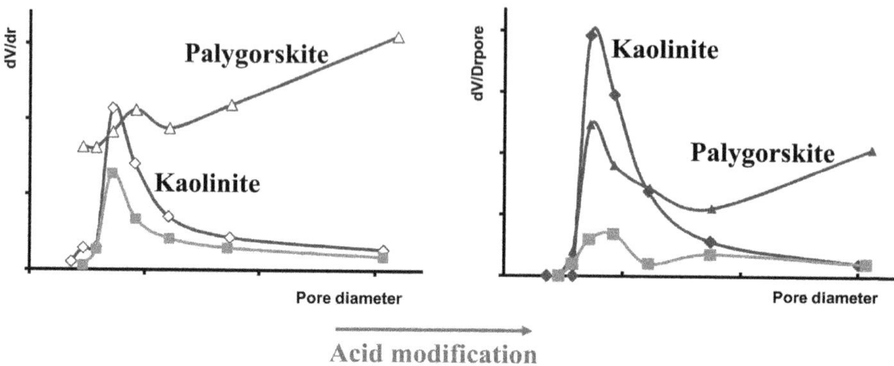

FIGURE 5.3   Increase in pore density through acid modification of typical aluminosilicates. Created, designed, and introduced by the author.

without a significant change in pore diameter sizes. The highest growth in mesopore density is attainable through the strongest acid attack on 1:1 layer-type structures, obviously due to the easy accessibility of protons to the octahedral sheet and octahedral aluminum cations. Structures in the octahedral sheet that are more resistant to acid modification are due to hidden tetrahedral sheets and stronger interactions between the layers. A typical acid modification increases SSA by 1.5 to 2.5 times without a significant change in pore size.

Alkaline or basic activation is employed as a modification method using alkalis, alkaline salts, alkaline earth metals, and oxides to control the alkaline (sodium) cation content. The alkaline treatment affects the tetrahedral sheets by chemically dissolving silica and alumina in the alkali, forming sodium silicates and sodium aluminates as follows:

$$SiO_2 + 2NaOH \rightarrow Na_2SiO_3 + H_2O$$

$$Al_2O_3 + 2NaOH \rightarrow 2Na_2AlO_2 + H_2O$$

In alkaline conditions, the Si–O–Si bond is less stable than the Si–O–Al bond, allowing silicon to easily transfer from the aluminosilicate structure into the solution. As a result, the silica content decreases, the alumina content remains unchanged, Na_2O content increases significantly (six or seven times), the magnesia content decreases drastically (three times), and several ion-exchange reactions occur, including the release of Mg (or Ca) ions from the interlayer space or octahedral positions and silanol groups, and the uptake of Na ions from the external solution, altering the theoretical form of the aluminosilicate to its alkaline form as follows:

$$x\mathrm{NaOH} + ((\mathrm{Si\text{-}O\text{-}Al})_n)^{x-} \cdot (x/2)\mathrm{Mg}^{2+} \rightarrow ((\mathrm{Si\text{-}O\text{-}Al})_n)^{x-} x\mathrm{Na}^+ + \left(\frac{x}{2}\right) \cdot \mathrm{Mg(OH)_2}$$

The role of NaOH solution can be summarized as disrupting the silicon skeleton of aluminosilicate by interrupting the -Si-O-Si-, -Si-O-Al-, and -Al-O-Al- bonds, followed by the formation of aluminate and silicate anions, as well as anions of silicon acids. The aluminate and silicate accumulated in the alkaline medium can trigger a polycondensation reaction between aluminate and silicate ions, either yielding an amorphous aluminosilicate phase or crystallizing the amorphous phase to form new aluminosilicate phases such as zeolite. In other cases involving combined aluminosilicate sorbents, alkali activation contributes to the transformation of the original aluminosilicate structure into a zeolite-like structure of heulandites with significant improvements in surface characteristics, including porosity, pore diameter, and specific surface area—approximately a 1.5 times increase in SSA, a 2.2-fold increase in total pore volume, and a 2.5-fold increase in mesopore volume. Such enhancements in the surface and structural elements of silicates are clearly evident in their ability to adsorb hydrogen through van der Waals' interactions.

Other processes, such as the melt-blending method, can incorporate organic cations into the interlayer space of aluminosilicate, making the hydrophilic surface more hydrophobic. The substitution of organic cations for exchangeable inorganic cations in aluminosilicate increases the interlayer distance and significantly enhances hydrogen adsorption capacity (HAC). Furthermore, combination techniques, such as using organic materials and biopolymers, metal-pillared aluminosilicate, or combinations with biochar or different aluminosilicates, along with thermal activation, can synthesize new aluminosilicate-based nanocomposite adsorbents. These modifications help to increase the SSA, adjust the pore size of clay minerals, and create various surface functional groups.

5.1.2 ZEOLITES

Zeolites, crystalline materials composed of SiO_4 or AlO_4 building blocks, contain intra-crystalline systems of channels (microporous aluminosilicates) and cages (5–20 Å) that can trap guest molecules such as hydrogen. The framework structures are relatively rigid and, like all microporous solids, have high SSAs and large pore volumes. Their properties can be controlled by altering the Si/Al ratio of the framework or their equivalent stoichiometry. The anionic nature of zeolite frameworks leads to the presence of cations within their structure. The main zeolite structures include sodalite SOD ($50NaO_2:Al_2O_3:5SiO_2$), zeolite LTA ($2NaO_2:Al_2O_3:1.926SiO_2$), and zeolite FAU ($16NaO_2:Al_2O_3:4SiO_2$). According to the International Zeolite Association (IZA) database, zeolites can be categorized into large pore aperture zeolites with a pore size of >7 Å (i.e., NaX, NaY, BEA, L-type, and ZSM-12), medium pore aperture zeolites with a pore size of 5–6 Å (i.e., ZSM-5, ZSM-11, ZSM-57, and silicalite), and small pore aperture zeolites with a pore size of 3–4 Å (i.e., RHO, DD3R, ZSM-58, SAPO-34, T-type, and AlPO-18) [2]. The typical unit cage, confined structure, and open porous 3D structure in zeolite adsorbents are shown in Figure 5.4. The theoretical hydrogen storage capacity of ~3 wt.% exists as an intrinsic geometric constraint of zeolites, with ~4.8 wt.% for sodalite (SOD) structures. Due to low theoretical gravimetric density, zeolites are not typically considered ideal

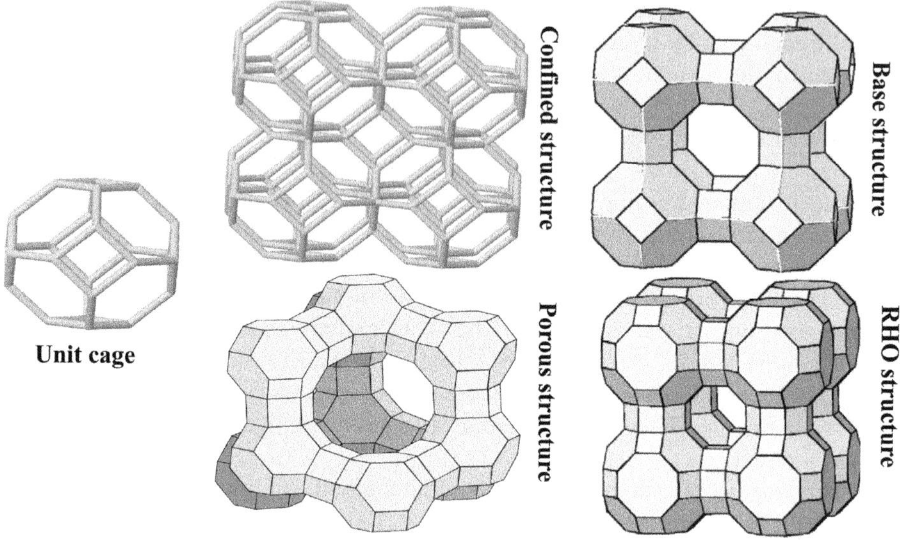

FIGURE 5.4 Schematic formation of unit cage, confined structure, and open porous 3D structure in zeolite adsorbents. Created, designed, and introduced by the author.

hydrogen storage materials. The isosteric heats are in the 6–7 kJ/mol range in typical hydrogen–zeolite systems. Zeolite structures can have intra-crystalline cavities on the order of the hydrogen molecule diameter. Meanwhile, their steric barriers function as molecular sieves, blocking the adsorption of larger gas molecules and exhibiting quantum-sieving effects on hydrogen isotopes. The heavy deuterium molecules can be adsorbed preferentially over the lighter hydrogen molecules due to their smaller zero-point motion. Other practical advantages over other microporous adsorbents include high thermal stability in comparison to metal–organic frameworks and organic polymers, high degassing temperature of the framework without risk of decomposition, easy characterization of the host material due to its crystalline nature, and well-defined pore sizes [3].

At room temperature and moderate pressure, hydrogen storage by zeolites is not significant. Hydrogen storage at higher temperatures has been proposed through the encapsulation mechanism in which hydrogen molecules enter and become entrapped in the sodalite cages. Enthalpies of adsorption, as a measure of the strength of the hydrogen–adsorbent interaction for zeolites, are in the 3.5–18 kJ/mol range. Hydrogen adsorption by zeolites focuses on the adsorption of gas at high temperatures, as hydrogen molecules rarely enter zeolite cages at room temperature and lower temperatures. At higher temperatures and pressures, hydrogen, with enhanced vibrational motion assisting access through the openings, can be forced into the cages. Subsequently, high temperatures on the adsorbent are required to release the trapped hydrogen in the zeolite cages. The application of zeolites as hydrogen storage systems in fuel cells requires conditions such as lower temperatures and high hydrogen storage capacity. The self-hydrogen storage capacities of different zeolite types include sodium-based zeolite X (1.79 wt.%), potassium-based zeolite X (1.96 wt.%), rubidium-based zeolite X (1.46 wt.%), cesium-based zeolite X (1.32 wt.%), magnesium-based zeolite X (1.62 wt.%), calcium-based zeolite X (2.19 wt.%), strontium-based zeolite X (1.68 wt.%), ultra-stable Y zeolite (0.4 wt.%), and zeolite Socony Mobil-5 (2.89 wt.%).

5.1.3 HYDROTALCITE

Hydrotalcite is generally known as a magnesium–aluminum hydroxycarbonate with a general formula $Mg_6Al_2CO_3(OH)_{16} \cdot 4(H_2O)$ presenting a layered crystal structure. Hydrotalcite, or hydrotalcite-like compounds (HTlcs), are mixed metal oxides and also layered double hydroxides. HTlcs are ionic and basic aluminosilicates composed of positively charged brucite ($Mg(OH)_2$) layers with trivalent cations substituting divalent cations at the centers of the octahedral sites of the hydroxide sheet. The hydroxide sheet contains hydroxide ions, where each –OH group is shared by three octahedral cations and points to the interlayer regions. A schematic representation of the hydrotalcite-like structure is given in Figure 5.5, with

$$A^{n-} : CO_3^{2-},\ SO_4^{2-},\ NO_3^-,\ Cl^-\ ,\ \text{and}\ M^{2+} - M^{3+} : Mg^{2+},\ Ni^{2+},\ Zn^{2+} - Al^{3+},\ Fe^{3+},\ Cr^{3+}$$

with

$$M^{2+}/M^{3+} = 2-4$$

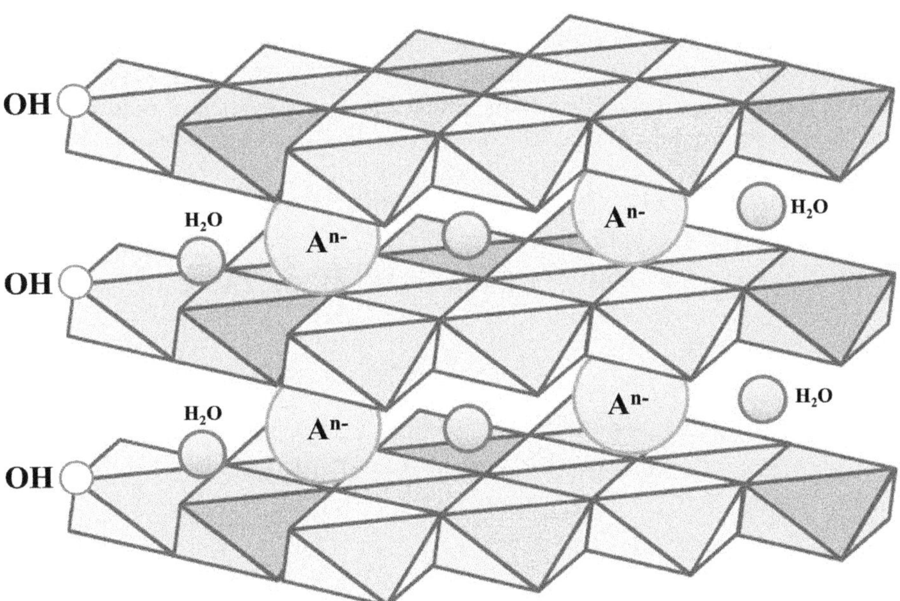

FIGURE 5.5 Typical hydrotalcite-like structure. Created, designed, and introduced by the author.

The main characteristics of HTlcs, such as permanent ion-exchange and adsorption capacity, mobility of interlayer anions, large surface areas, and the stability and homogeneity of the material formed by thermal decomposition, have made them applicable candidates for hydrogen adsorption in synthesized types. The synthesis of HTlcs can be carried out via a wide range of compositions and materials with various M(II)/M(III) cation combinations, as well as M(I)/M(III) cation pairs with different anions. The preparation methods for HTlcs include co-precipitation, hydrothermal methods, urea hydrolysis, and the sol–gel method. The physicochemical properties, such as crystallinity and surface area, can be defined through the manufacturing stage.

Hydrotalcite-like compounds can be used to adsorb gases via the naturally occurring hydrotalcite material, which has the best overall adsorption capacity and kinetics. Nevertheless, the hydrotalcite material prepared by high supersaturation methods has shown high gas adsorption compared to other oxide adsorbents such as alumina and magnesia due to its high surface area and basicity. $Mg^{2+}Al^{3+}(Fe(CN)_6)^{4-}$ is considered a high-capacity gas adsorbent. Nanocrystalline hydrotalcite-derived reduced mixed oxides containing magnesium, nickel, and aluminum exhibit a 3.9 wt.% HAC, with the hydrogen adsorption enthalpy and entropy changes of the reduced mixed oxides being 47.58 kJ/mol and 120.98 kJ/mol K, respectively, along with a desorption activation energy of 65 kJ/mol, corresponding to their potential as energy storage materials. The pore diameter is estimated to be in the 40–70 Å range, with SSAs ranging from 186 to 230 m^2/g. The surface area and pore volume of the hydrotalcite substrates are directly related to the added metal ions, as they decrease with increasing Co content, while the average pore size between the crystallites increases. In contrast, the presence of Ni in Ni-Mg-Al does not affect the SSA and is more dependent on aluminum content, as the specific surface area decreases with an increment of aluminum. The increase in gallium metal content in the Mg-Ga-Al HTlcs decreases the surface area of the calcined material. The Ni-Mg-Al HTlcs with Mg experience reduction during hydrogen adsorption, which causes segregation of both magnesia and NiO phases from the HTlcs. The overall reaction is given as follows:

$$(M^{2+}_{1-x}M^{3+}_x(OH)_2)(A^{n-})_{x/n} \cdot mH_2O \rightarrow (M^{2+}_{1-x}M^{3+}_x(OH)_2)^{x+}(A^{n-})_{x/n}$$

$$(M^{2+}_{1-x}M^{3+}_x(OH)_2)^{x+}(A^{n-})_{x/n} \rightarrow (M^{2+}_{1-x}M^{3+}_xO)^{x+}(A^{n-})_{x/n}$$

$$M^{2+}_{1-x}M^{3+}_xO)^{x+}(A^{n-})_{x/n} \rightarrow M^{2+}_{1-x}M^{3+}_xO_{1+(x/2)}$$

$$M^{2+}_{1-x}M^{3+}_xO_{1+(x/2)} \rightarrow M^{2+}O + M^{2+}M^{3+}O_4$$

with

$$MgO + H_2 \leftrightarrow Mg^0 + H_2O$$

$$NiO + H_2 \leftrightarrow Ni^0 + H_2O$$

5.1.4 CLAY SPECIES

As an aluminosilicate, clays are made from silicon–oxygen tetrahedrons and aluminum–oxygen octahedrons. The layer thicknesses and interlayer spacings of typical clay minerals are approximately 1 nm, a space that can be used for hydrogen adsorption. The thickness of a typical clay layer is ~9 Å, and the interlayer space is ~12 Å. Due to their unique molecular structures and large surface areas, clays exhibit excellent HAC. The combination of illite, smectite, and kaolinite can adsorb about 500 ppm (0.25 mol/kg) of hydrogen. Sudoite (Al–Mg dioctahedral smectite) is also a good adsorbent for hydrogen storage. Montmorillonite (MMT) exposure to pure hydrogen gas at low temperatures up to ambient temperature, with increasing pressure up to 50 bar, results in significant hydrogen adsorption (~2 cm^3/g). At all temperatures, hydrogen adsorption increases non-linearly with pressure and reaches a saturation plateau. The Langmuir and Freundlich isotherms can explain the adsorption behavior (small least-squares error). Other clay hydrogen adsorbents include MMT, illite, kaolinite, smectite, sepiolite, and chlorite. The sepiolite and smectite structures contain significantly higher excess adsorption than MMT or illite, while kaolinite and chlorite exhibit low or negligible excess adsorption. Kaolinite and talc, due to a lack of layer charges, cannot attract interlayer cations and have no binding sites for hydrogen adsorption. In illite, potassium ions are the main interlayer cations, and the interlayer space is too small for hydrogen gas, which limits its adsorption. The differences in HAC among different clay minerals can be attributed to variations in their pore structures, including specific surface areas and micro/mesopore volumes. Additionally, more complex pore structures and larger specific surface areas provide more adsorption sites for hydrogen molecules. Nevertheless, HAC increases with increasing pressure and decreases with increasing temperature in all clay minerals, such as mica. The one-dimensional porous phyllosilicate minerals, such as sepiolite and palygorskite, with their 2:1 chain-like fibrous structure, possess high specific surface areas, relatively high porosity, and multiple surface charges, which result in significant adsorption performance and hydrogen storage capacity. MMT particles often take a cellular form with enveloped filamentous or wavy bumps and are interconnected. The edges of pores exhibit leaf-like and petal-like layered structures, forming wide and narrow fissure-like connections deep inside. Additionally, honeycomb-shaped pores may develop with pore diameters less than 50 nm, creating favorable conditions for hydrogen adsorption.

Sepiolite and palygorskite belong to hydrated chain-layered magnesium silicates of phyllosilicates with a 2:1 chain-like structure and fibrous morphology, consisting of needle-like or belt-like crystals measuring 1–10 μm in length and 0.01 μm in width. They have characteristics similar to zeolites, such as high specific surface areas, well-developed pores, and polar surface features, which make them suitable for hydrogen adsorption. Due to the periodic inversion of the tetrahedral active oxygen, the octahedral sheets are discontinuous and form unique one-dimensional channels that extend along the c-axis. These channels are highly regular, equal in size, and parallel but non-interacting, creating a honeycomb appearance in cross section. As an adsorption advantage, the cross-sectional area of individual channels in palygorskite and sepiolite is larger than the kinetic diameter of a hydrogen molecule. Owing to the regular shape of the channels, palygorskite and sepiolite allow for hydrogen molecule adsorption through channel entry as well as onto the surfaces outside the channels. The diameters of the irregular spheres range from 2 to 8 µm. Numerous small fiber particles with diameters in the nanometer range aggregate into irregular pores, providing various sites for hydrogen adsorption.

Besides specific surface area, pore size, and pore structure, the surface adsorption potential can significantly influence the surface adsorption of hydrogen molecules. Surface functional groups serve as centers of residual charges, which make them active sites for the adsorption of polar adsorbate molecules or easily polarizable molecules (e.g., hydrogen). Palygorskite and sepiolite can create different types of surface functional groups, including (i) tetrahedral siloxane holes on tetrahedral sheets, (ii) magnesium alcohols on octahedral sheets, (iii) silicon and magnesium alcohols on sheet edges, and (iv) Lewis acid sites on edges. Therefore, palygorskite and sepiolite act as highly polar adsorbents with a relatively high density of surface functional groups, effectively adsorbing cations for polar or easily polarizable molecules. Due to the structural stability of most aluminosilicates, contamination of adsorbed hydrogen is unlikely, even when used under low-temperature conditions, benefiting from fast charge–discharge kinetics and long-term reversible cycling.

Mechanisms of hydrogen adsorption in porous aluminosilicates follow the solid-state hydrogen storage mechanisms classified as chemical or physical storage. The potential energy depicting the interaction between a hydrogen molecule and a solid aluminosilicate surface is shown in Figure 5.6. For hydrogen, the potential energy curve near a solid surface exhibits two energy minimum points at different depths; the one closer to the adsorbent represents physical adsorption at an equilibrium distance farther from the adsorbent, whereas the other at a closer distance corresponds to chemical adsorption of hydrogen. Physical adsorption occurs in most

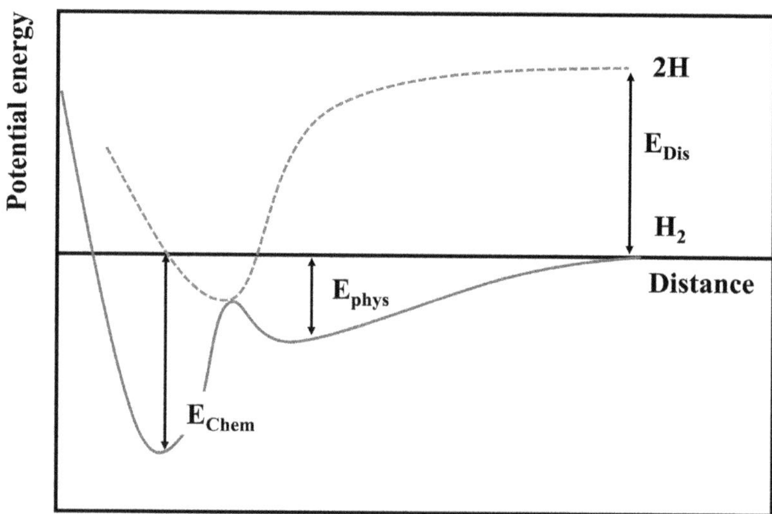

FIGURE 5.6 E_{chem} (potential energy for chemical adsorption) and E_{phys} (potential energy for physical adsorption) in the potential energy versus distance curve for a solid-state adsorbent surface. Created, designed, and introduced by the author.

porous aluminosilicates, where hydrogen is adsorbed as a monolayer at temperatures above the boiling point. The combined effects form a shallow potential energy curve minimum at one molecular radius from the surface. The energy of second-layer adsorption is related to the vaporization enthalpy of the adsorbate. Therefore, at pressures and temperatures above the boiling point of the adsorbent, the adsorption process leads to the formation of a relatively stable adsorption layer with gas molecules adsorbed in a monolayer on the surface of the solid aluminosilicate. Physical adsorption in porous aluminosilicates offers the advantages of complete reversibility and fast kinetics compared to chemical storage in compounds. Porous aluminosilicates can adsorb hydrogen either by providing a high surface area or by encapsulating or capturing hydrogen in microporous media.

Physical adsorption of hydrogen involves charge polarization, wherein the charged ions in the substrate generate a local electric field to polarize hydrogen, with the generated electric field depending on the charge of the metal ion. For instance, the positive charges of Mg and Al in palygorskite are 2.0–2.4, while the positive charge of Mg in sepiolite is 2.0. The hydrogen molecules interact with a side-pair configuration relative to the metal ion site at a distance greater than 5 Å. As a result, adsorbed hydrogens in palygorskite and sepiolite are located on the outer surface and in the bulk phase, while montmorillonite and chlorite can be adsorbed only on the outer surface. The hydrogen storage characteristics of aluminosilicates are summarized in Table 5.3.

5.1.5 Metal Oxides and Mixed Metal Oxides

The single-component metal oxides (MOs) and mixed metal oxides (MMOs) are promising candidates for hydrogen energy storage systems, particularly for applications such as gas storage through physisorption of hydrogen. Adsorption-enhancing

TABLE 5.3

Hydrogen Storage Characteristics of Typical Aluminosilicates

Adsorption isotherms	Langmuir
Adsorption capacity	0.5–18.5 cm³/g
Specific surface area	18–230 m²/g
Micropore volume	$4.3–100 \times 10^{-3}$ cm³/g
Micropore specific surface area	21–301 m²/g
Mesopore volume	$40–68 \times 10^{-3}$ cm³/g
Advantage	High cycling
	Easy regeneration
	Thermal stability
	Chemical stability
	Non-toxicity and availability
Disadvantage	Structure dependency
	Contamination sensitivity

mechanisms can include the application of electric fields or doping with redox additives. In conductive MOs and MMOs, the externally applied electric field affects the dipole moment, activating hydrogen polarization along the surface and causing stronger adsorption of hydrogen molecules. Additionally, the presence of redox species with multiple oxidation states can enhance the overall hydrogen storage capacities. The variety of MMOs containing alkali, alkaline, rare earth, and noble metals can be considered active hydrogen storage materials. Their activity is related to their ability to dissociate hydrogen molecules into hydrogen atoms as a step for adsorption/absorption. The hydrogen storage capacity is affected by their physical and chemical properties, mechanisms, and efficiencies in adsorption [4].

Oxide, a chemical compound containing at least one oxygen atom and one other element in its chemical formula (metal and metalloid actions), and MMOs, an oxide containing two or more metal and metalloid cations, are the bases for many advanced ceramics including complex MOs (MOs with the presence of multiple metal/metalloid cations). While MOs are simple structures, MMOs can be formed in binary, ternary, quaternary, and more complex structures. Most MOs are crystallized (otherwise they can be considered glass formers), whereas MMOs can be either crystalline or amorphous, which is determined based on their oxide compositions and crystalline phase content. The general form of oxide structures is based on M_xO_y with x and y factors depending on cation valency capacity. The crystalline MMO compounds can be formed in perovskites, scheelites, spinels (AB_2O_4), poly-crystal ($A_xB_yO_z$), and garnets ($A_3B_5O_{12}$) arrangement with possessed the general formulas of ABO_3, ABO_4, AB_2O_4, $A_3B_2O_8$, and $A_3B_5O_{12}$. The exact arrangement of the metal/metalloid cations may differ from their theoretical coordination and the arrangements of other cationic species in similar structures. It is crucial to determine which metal cations should be considered as the active centers of MMO structures. The significant role of MMOs as hydrogen adsorbents is related to their structure, acid/base affinity, and redox properties. MOs and MMOs can be formed in nanostructured forms, including nanoparticles, nanocages, nanotubes, nanofibers, and even 2D building blocks and nanosheets (e.g., zinc oxide and aluminum-doped zinc oxide or perovskite-type $Ca_2Nb_3O_{10}$ nanosheets). The nanoconfinement effects of MMOs lead to complex structures with high adsorption surface areas, with the incorporated molecules crystallized in two stages of nucleation and growth. The defects and vacant sites, along with misplaced atoms, are non-stoichiometric features of MMOs, representing disorders in the crystal formation and chemical formula of the MMOs. While these imperfections are thermodynamically favorable (improving chemical reactivity), they can also enhance the kinetics of adsorption. The formed defects in MMOs can be categorized as intrinsic (in the pure substance) and extrinsic defects (the presence of impurities) and are responsible for the enhancement of solid-state hydrogen storage. For instance, a perovskite-type MMO with stable chemical bonds can form connections between protons and oxygen in the oxide (substitutional OH ion defects), which can be amplified by the vacancies in the perovskite as adsorption sites for protons. Therefore, the physical properties of MMOs are associated with the structures and energy of ionic and semi-covalent bonds in solids, making it critical to consider the interactions between ions. The electronic structures of MMOs, as indicators of mutual electron and ion interactions, can be indicative of hydrogen adsorption affinity. For example, in semiconducting MMOs, hydrogen tends to bond covalently with surface atoms based on the atomic configuration of the semiconductor. The curvature in the surface morphology of the nanoparticles facilitates the absorption/desorption of hydrogen. It should be noted that the bulk surfaces of some MMOs are metastable for high H-content, and as hydrogen is electrically active, the H atoms (H^+ and H^-) adsorb to regions of high or low electronic charge density, where they can interact with anions or cations on MMOs surfaces. Hydrogen adsorption on sites of oxide clusters (MOs) meets the adsorption energy criteria, with bond strengths ranging from 0.15 to 0.21 eV. In an MOs system (M = Mg, Ba, Be), the energy profiles and kinetic constants for the hydrogen molecule splitting reaction facilitate hydrogen adsorption on their sites, with rapid uptake/release at operating temperatures and pressures.

Detailed information on the MO and MMO systems with high storage potential is as follows. *Storage capacity* (*SC*) can be estimated using charge/discharge current (*I*-mA), discharge time (t_d hours), and active mass (*m*-geram) via $SC = It_d / m$ equation using data derived from the charge/discharge curves of the electrodes. As the generated double layer charging process and faradaic reaction exceed the equilibrium potential, hydrogen can be absorbed/stored at the surface of the MMOs.

5.1.5.1 Cerium Vanadate (CeVO$_4$) Nanoparticles

Cerium vanadate, a member of the rare earth orthovanadate nanomaterial family identified as vanadates (AVO_4) (e.g., bismuth vanadate $BiVO_4$, iron vanadate ($FeVO_4$), gadolinium vanadate ($GdVO_4$), lanthanum ($LaVO_4$), indium vanadate ($InVO_4$), and samarium vanadate ($SmVO_4$)), is an orthovanadate composed of cerium and vanadium oxides. Other vanadium-based compounds include vanadium as a central ion in transition metal vanadium oxides (TMVOs), which are used in the production of rechargeable adsorbents with cation-exchange potential, enhanced electrical and optical properties, and energy storage potential. As vanadium (V) has five valence electrons, it can adopt multiple oxidation states; this variation in oxidation states can be reflected in the properties of MMOs containing V ions. The $CeVO_4$ unit cell structure belongs to the space group I41/amd and has the crystal structure of tetragonal zircon (4f electronic structure) with three additional distinct polymorphic forms [5]. The $CeVO_4$ NPs can be easily manipulated in size, shape, and surface area through techniques like hydrothermal synthesis, microwave synthesis, co-precipitation, combustion, and sono-chemical methods (using ammonium metavanadate, cerium (III) nitrate hexahydrate as the primary reactant, and hydrazine as the source of OH) by altering parameters including temperature, pH, solvent, concentration, and surfactants. The applied sono-chemical method utilizes acoustic cavitation, which helps bubble collapse and produces intense local heating and high pressures. It is an accessible, chemically stable, non-toxic, and highly photocatalytic active oxide (with a bandgap of 2.8 eV. $CeVO_4$ nanoparticles exhibit adsorption/desorption isotherms of type IV with H3-shaped hysteresis loops, indicating mesoporous structures (porous spherical nanostructures with a size range of 5–30 nm) [6]. NPs can be manufactured with a specific area as high as 40–110 m^2/g, a total pore volume of 0.24 cm^3/g, and average pore diameters of 3–20 nm , resulting in high effectiveness for hydrogen SC. According to electrochemical hydrogen storage evaluations (hydrogen storage on inorganic compounds), after 15 consecutive discharge sequences, the tendency of nanoparticles for hydrogen sorption (in a solid-state matrix) can be estimated to be ~4300 mAh/g, with different potential plateaus in discharge patterns. Results are related to the progression of the following reactions:

$$z e^- + CeVO_4 + z H_2O \rightarrow z OH^- + z H + CeVO_4$$

$$z H + CeVO_4 + z OH^- \rightarrow z e^- + CeVO_4 + z H_2O$$

The electrochemical performance of vanadium-based compounds involves multistep vanadium reduction/oxidation reactions. For instance, ZnV_2O_4 glomerulus nano/microspheres have shown adsorption/desorption profiles indicative of hydrogen storage potential. The hydrogen adsorption and interactions are attributed to the more oriented lattice planes in the crystal structure after hydrogen adsorption. Furthermore, hydrogen adsorption can reduce defects and vacancies in ZnV_2O_4 nano systems. Hydrogen storage of about 2.2 wt.% at high temperatures (>200 °C) with fast adsorption (12 min to maximum capacity) and desorption (83% in less than 30 min) is achievable, with the apparent activation energy calculated to be 50 kJ/mol.

Vanadium oxide (vanadia, V_2O_5) can significantly improve hydrogenation/dehydrogenation temperature and the reversible hydrogen storage properties as it affects the electrochemical properties of the adsorbents. The combination of M-V-Os or V-Os with any active substrates can enhance hydrogen storage performance through the occurrence of multilevel hydrogen sorption[7]. The V_2O_5 presence of these compounds in adsorbent structures can improve total adsorbed hydrogen by up to ~100% and decrease the apparent activation energy (E_a) to half. V_2O_5 can act as catalysts and reduce the reaction barrier of the hydrogenation/dehydrogenation process, thereby resulting in improved reversible hydrogenation and dehydrogenation (Figures 5.7 and 5.8).

5.1.5.2 Aluminum Mixed Oxides

Aluminum, as one of the most important metal ions in the structure of hydrogen storage materials like MMOs, creates different adsorbent compounds including nanosystems of $ZnAl_2O_4$, $BaAl_2O_4$, $NiAl_2O_4$, $CoAl_2O_4$, $Sr_3Al_2O_6$, and $Dy_3Al_2(AlO_4)_3$ with high discharge capacity and hydrogen content. The more complex structures of garnets and polycrystals, compared to aluminum mixed oxides (MalOs), show higher discharge capacities. Forming composite and hybrid structures with MAlOs leads to higher SSA, more activation of redox species, and enhanced multilevel hydrogen storage affinity.

5.1.5.3 $ZnAl_2O_4$ and $ZnAl_2O_4$-TiO_2

Zinc aluminate ($ZnAl_2O_4$), a compound that can be formed by the combination of alumina and zinc oxide as a ternary oxide, crystallizes in a spinel structure. Its natural form, "gahnite," consists of all zinc atoms in tetrahedral sites and all aluminum ions in octahedral sites of the FCC lattice of oxygen atoms. Zinc aluminate can be

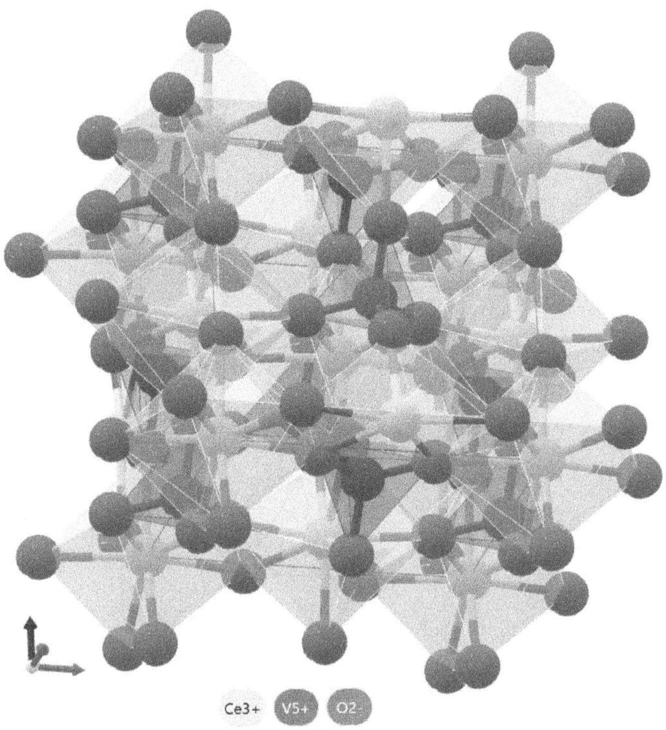

FIGURE 5.7 Crystal structure and ions positions of cerium vanadate. Created, designed, and introduced by the author.

manufactured through solid-state reaction, chemical combustion, sol–gel, hydrothermal, solvothermal, and co-precipitation methods. Via co-precipitation-assisted hydrothermal routes, aqueous solutions of aluminum and zinc nitrates with tetraethyl orthotitanate (TEOT -$(C_2H_5O)_4$Ti) can be co-calcined (~900°C) to synthesize $ZnAl_2O_4$-TiO_2 NPs. The adsorption process in the electrochemical aqueous electrolyte can be summarized as follows:

$$ZnAl_2O_4 + xH_2O + xe^- \leftrightarrow (ZnAl_2O_4 + xH) + xOH^-$$

The reaction during the charging process indicates how hydrogen atoms (H) are produced throughout the electrochemical decomposition in the solution, migrate to the adsorbent electrode, and get stored. The opposite direction is the discharging reaction, where H atoms are released from the adsorbent electrode under alkaline conditions. The SC is accompanied by meso- and macroporous structures, showing a type III isotherm with an H3 hysteresis loop [8]. This behavioral isotherm is typically related to solids containing agglomerates or aggregates of slit-shaped pore particles (edged or plate-like particles such as cubes) with irregular shapes and/ or size. An SSA of ~14.5 m^2/g, a pore size of nearly 15.5 nm, and a pore volume as

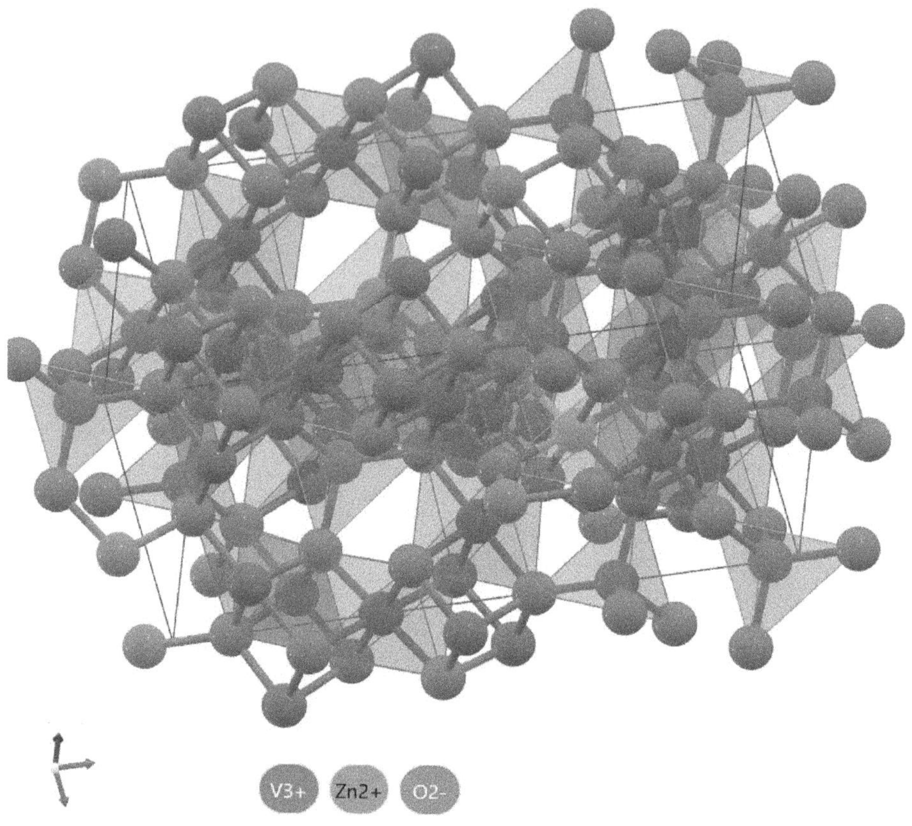

FIGURE 5.8   Crystal structure and ions positions of zinc vanadate. Created, designed, and introduced by the author.

low as $0.05\,cm^3/g$ can be achieved. The presence of a second oxide, such as titania, shifts pore sizes to a smaller range of ~10.5 nm, with a higher SSA of $48\,m^2/g$ and an increased pore volume of $0.16\,m^3/g$. The discharge capacity of base $ZnAl_2O_4$ remains nearly twice $ZnAl_2O_4$-TiO_2 over 15 adsorption cycles, with a similar discharge plateau potential (Figure 5.9).

5.1.5.4 Zn_2GeO_4

Zinc germanate oxide with Zn_2GeO_4 stoichiometry crystallizes in the trigonal R3 space group, with two inequivalent Zn divalent sites. Zn should be bonded to four O atoms to form ZnO_4 tetrahedra that share corners with four equivalents GeO_4 tetrahedra. The O sites include bonding in a trigonal planar geometry to two equivalent Zn atoms and one Ge tetravalent atom. Through a chemical precipitation method involving zinc nitrate and germanium tetrachloride solution as Zn and Ge precursors and acacen as a capping agent, Zn_2GeO_4 nanostructures could be synthesized after calcination at 1000°C. Adding a black graphene suspension and employing a pre-graphenization technique leads to the formation of Zn_2GeO_4/ graphene

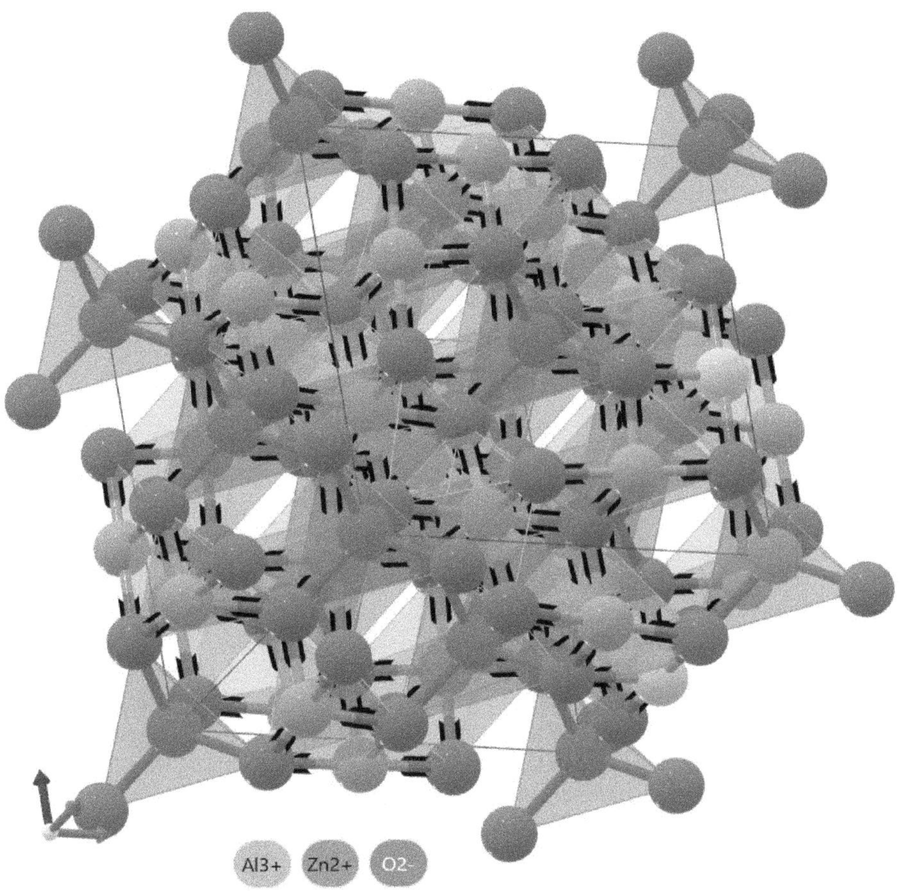

FIGURE 5.9 Crystal structure and ions positions of zinc aluminate. Created, designed, and introduced by the author.

nanocomposites with nanoparticles adsorbed on the graphene surface. The adsorption/desorption isotherms of nanoparticles match type IV isotherms with H2-type hysteresis, a characteristic of solid particles connected by nearly cylindrical channels or made by aggregates (consolidated) or agglomerates (unconsolidated) of spheroidal particles with pores of non-uniform size or shape (type H2) [9]. While Zn_2GeO_4 NPs reach 800 mAh/g (~2.83 wt.%) after 25 cycles, Zn_2GeO_4/graphene nanocomposites show hydrogen capacities around 2700 mAh/g (~9.54 wt.%) after several cycles. The surface quality of Zn_2GeO_4 NPs (8–40 nm) include a total pore volume of ~0.22 m^3/g and a mean pore diameter of ~2nm, with an SSA of ~220 m^2/g (Figure 5.10).

5.1.5.5 $BaAl_2O_4$

The compounds with a hexagonal structure (space group P63), such as $BaAl_2O_4$, consist of Ba^{2+} ions situated within channels of AlO_4 tetrahedra. Spinel $BaAl_2O_4$ NPs (30–40 nm) can be synthesized via aqueous combustion using stoichiometric

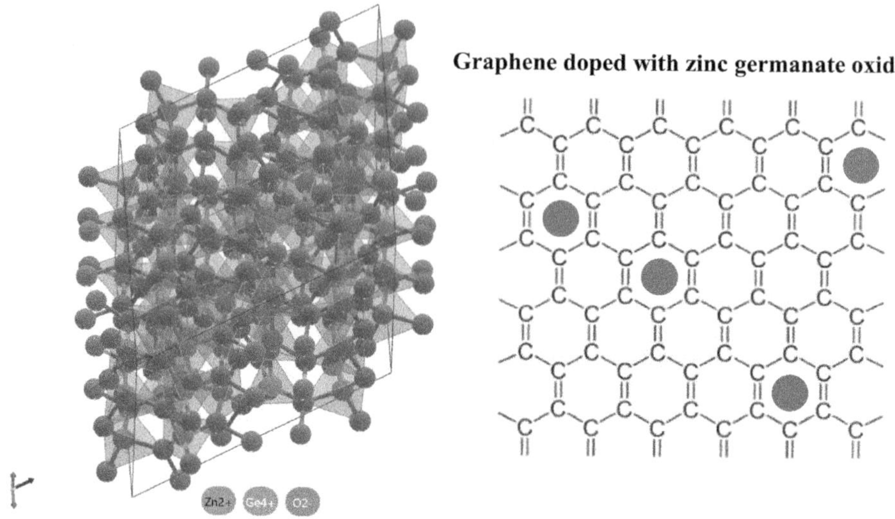

FIGURE 5.10   Crystal structure and ions positions of zinc germanate. Created, designed, and introduced by the author.

amounts of Ba and Al cations in aqueous barium nitrate and aluminum nitrate nonahydrate solutions with additional fuel. The discharge capacity during 15 electrochemical cycles reaches 1000 mAh/g [10]. The hydrogen interaction in an aqueous electrochemical cell indicates a hydrogen adsorption mechanism. The charging process occurs on electrodes via an oxidation process [11]. During electrolyte dissociation, hydrogen migrates from the electrolyte and is absorbed by the adsorbent surface, with an electron formed in the opposite direction, which can be explained as follows (Figure 5.11):

$$BaAl_2O_4 + xH_2O + xe^- \leftrightarrow (BaAl_2O_4 + xH) + xOH^-$$

5.1.5.6   NiAl$_2$O$_4$

As NiAl$_2$O$_4$ is a binary oxide system (NiO $-$ Al$_2$O$_3$), it can be synthesized to MMO through thermal decomposition and auto-combustion to form a monolith with composite NPs (with rutile, silica, or graphene. The NiAl$_2$O$_4$ spinel forms as a double oxide with a normal spinel-type structure, where Al ions occupy octahedral positions and Ni ions are accommodated in tetrahedral positions. According to this structure, there are four octahedral positions and eight tetrahedral positions where hydrogen can be adsorbed in the cavities of the tetrahedral and octahedral sites in the spinel [12]. The synthesized powders are nanoporous (7–20 nm pores) with enhanced electrochemical hydrogen storage properties, achieving an SSA of up to 35 m^2/g and an increased pore volume of 0.13 m^3/g. The discharge capacity is nearly 1000 mAh/g after 15 electrochemical cycles (~3.2 wt.%) in pure forms, which increases to nearly

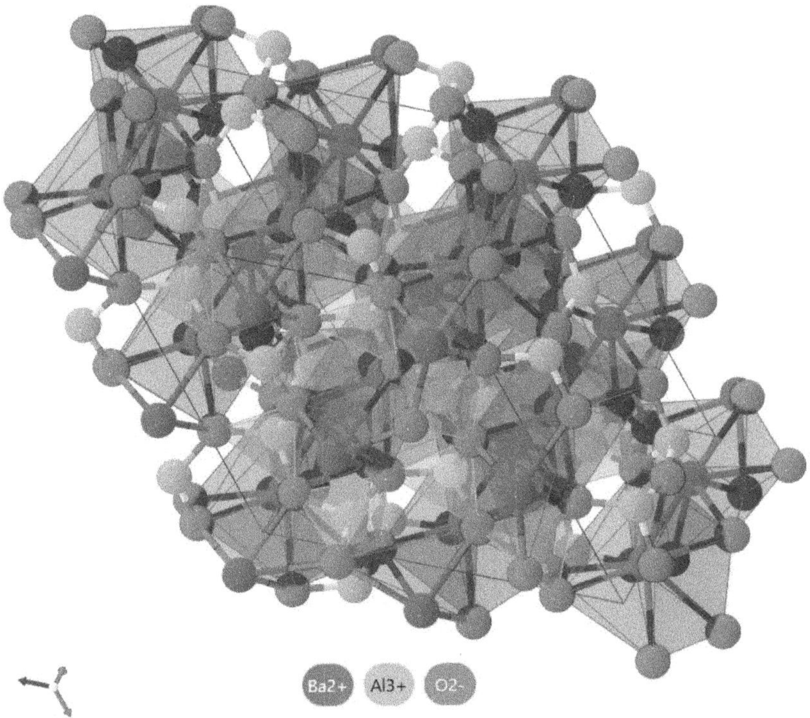

FIGURE 5.11 Crystal structure and ions positions of barium aluminate. Created, designed, and introduced by the author.

three times in composite NPs with extra NiO and rutile additives (~9.8 wt.%) [13]. The plateau of discharge potentials is in the 2–10 V range. The interaction reaction can be as follows:

$$NiAl_2O_4 + xH_2O + xe^- \leftrightarrow (BaAl_2O_4 + xH) + xOH^-$$

The H/M ratio (number of hydrogens to the molar mass of the material host) can be estimated to be 0.85 and 0.93 in pure and composite NPs. The lower hydrogen sorption sites generated on the surface due to inappropriate additives, such as silica, decrease adsorption capacity. The type III isotherm with a large H3-type hysteresis loop is typically reported for this system, indicating the involvement of aggregates or agglomerates of particles forming slit-shaped pores with a non-uniform size and/ or shape (Figure 5.12).

5.1.5.7 CoAl₂O₄

Cobalt aluminate (Thenard's blue, $CoAl_2O_4$) takes the structure of a general spinel with a cubic system and space group Fd3m, with the unit cell containing 32 O ions, 8 divalent cobalt cations occupying 8 tetrahedral sites, and 16 trivalent aluminum

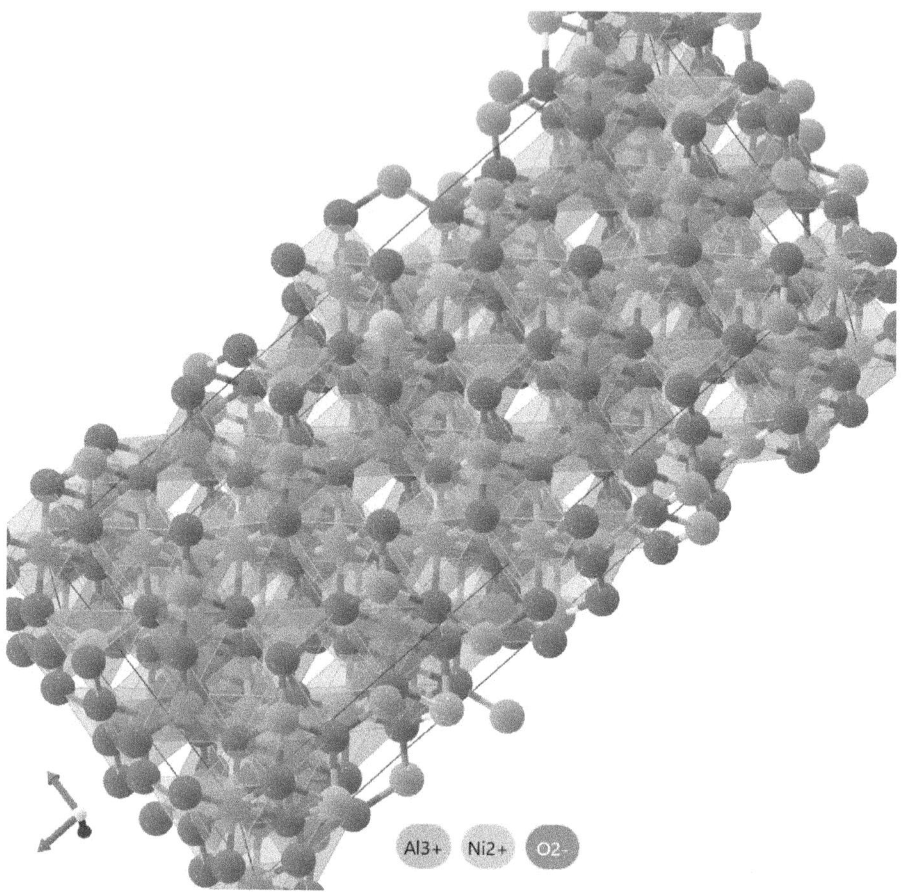

FIGURE 5.12 Crystal structure and ions positions of nickel aluminate. Created, designed, and introduced by the author.

ions occupying 16 octahedral sites. It can be synthesized via the solid-state reaction method using high-purity CoO and alumina calcined at 1200°C, or the thermal decomposition method with aluminum and cobalt nitrate nonahydrate in an alcohol–water solution calcined at 800°C to form 20 to 100 nm NPs [14]. The resulting powders have a capacity of 1100 mAh/g after 20 cycles, with two conjugated plateaus at approximately 0.1 V during discharging (Figure 5.13).

5.1.5.8 $NiCo_2O_4$

Nickel cobaltite, $NiCo_2O_4$, is a spinel-derived structure that crystallizes in the orthorhombic Imma space group. Divalent Co cations are bonded to four O atoms to form CoO_4 tetrahedra that share corners with six equivalent CoO_6 and NiO_6 octahedra. Trivalent Co cations can be bonded to six O atoms to form CoO_6 octahedra that share corners with six equivalent CoO_4 tetrahedra, edges with two equivalent

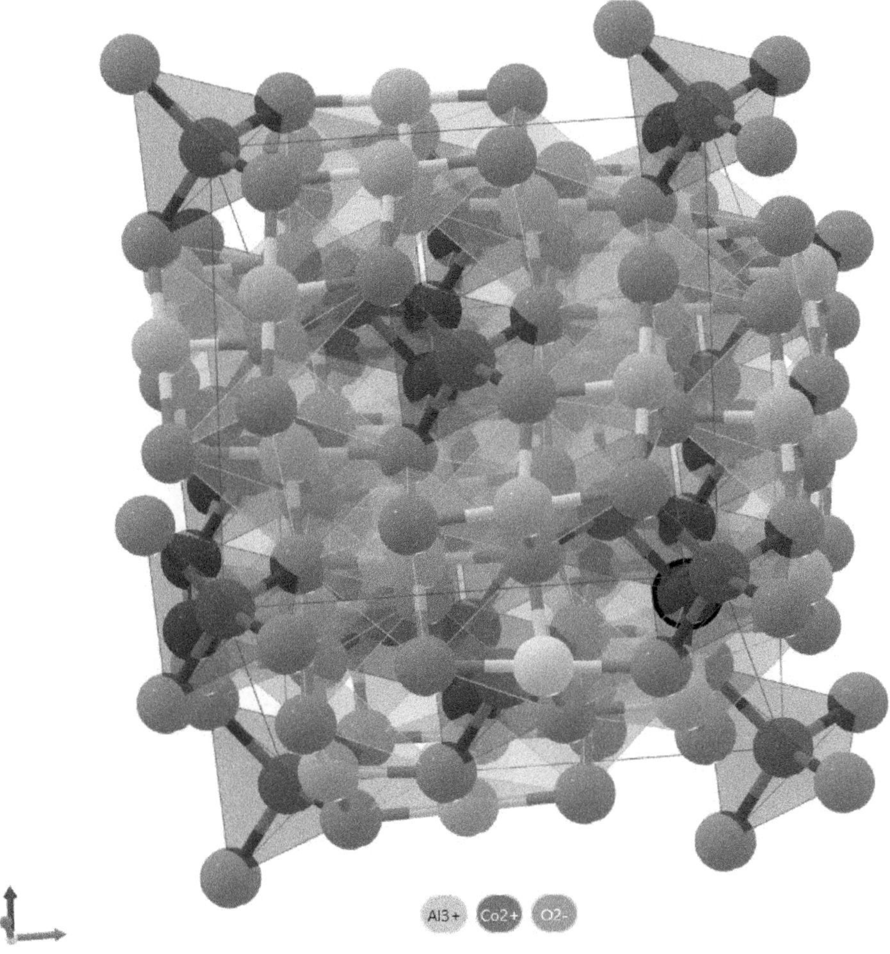

FIGURE 5.13 Crystal structure and ions positions of cobalt aluminate. Created, designed, and introduced by the author.

CoO_6 octahedra, and edges with four equivalent NiO_6 octahedra. Ni trivalent cations can be bonded to six O atoms to form NiO_6 octahedra that share corners with six equivalent CoO_4 tetrahedra, edges with two equivalent NiO_6 octahedra, and edges with four equivalent CoO_6 octahedra. $NiCo_2O_4$ NPs can be synthesized using a sono-chemical method using nickel(II) bis(acetylacetonate) and cobalt(II) acetate as precursors, followed by calcination at 900°C [15]. Although $NiCo_2O_4$ NPs as hydrogen storage material show more than 2000 mAh/g discharge capacity (related to the reaction below and the redox behavior of both nickel and cobalt), additives such as rutile can further improve the capacity to more than 2500 mAh/g after ten cycles.

$$NiCo_2O_4 + xH_2O + xe^- \leftrightarrow (NiCo_2O_4 + xH) + xOH^-$$

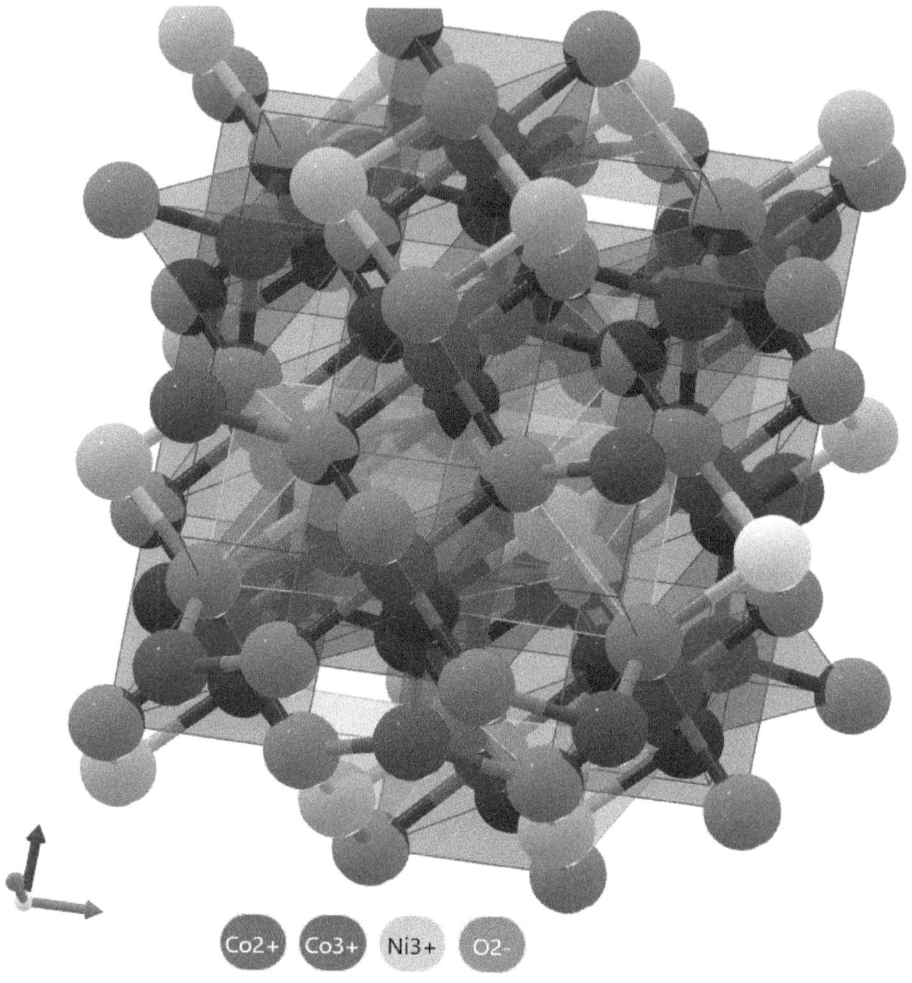

FIGURE 5.14 Crystal structure and ions positions of nickel cobaltite. Created, designed, and introduced by the author.

Nanoparticles with a discharge plateau potential of 0.35 V exhibit good electrochemical hydrogen storage performance (Figure 5.14).

5.1.5.9 $Dy_3Fe_5O_{12}$ and $DyFeO_3$

Dysprosium–iron oxide ($DyFeO_3$) is a binary metal oxide (BMO) with an orthorhombic perovskite structure that crystallizes in the orthorhombic Pnma space group. Dy trivalent cations are bonded in an eight-coordinate geometry to eight O atoms. Fe trivalent cations are bonded to six O atoms to form corner-sharing FeO_6 octahedra with an octahedral tilt angle of 37°–38°. It is a representative of ABO_3 stoichiometry adapted from $M_3M'_2(XO_4)_3$ by M and M' di- and trivalent cations with garnet structures such as $Dy_3Fe_5O_{12}$. Using solution combustion of hydrated nitrate precursors with carbohydrate fuel calcined at 750°C, particle sizes in the range of 10–25 nm can be synthesized. The electrochemical charging/discharging processes for $y_3Fe_5O_{12}$ and $DyFeO_3$ include (i) the first layer on the surface of the particles over a few atomic layers of the materials and (ii) the final adsorption inside the particles with the diffusion of H atoms. The discharge capacities of 2000 and 2100 mAh/g after 15 electrochemical cycles are recorded for $Dy_3Fe_5O_{12}$ and $DyFeO_3$, respectively [16]. The $DyFeO_3$ NPs show higher discharge capacity. The hydrogen storage in these MMOs occurs through a series of distinct stages in the electrochemical process with different potential plateaus in the charge and discharge profiles. This facilitates various distinct sites for hydrogen sorption on the surface and into the nanopores of the bulk material. The significant discharge capacity profiles for both oxides indicate their suitability for H-sorption, according to the following equation:

$$Dy_3Fe_5O_{12}/DyFeO_3 + xH_2O + xe^- \leftrightarrow \left(Dy_3Fe_5O_{12}/DyFeO_3 + xH\right) + xOH^-$$

The discharge pattern of the $Dy_3Fe_5O_{12}$ NPs contains triple flat potential plateaus, which can be attributed to special structural defects, roughness, and bonding. Meanwhile, the discharge curves of $DyFeO_3$ NPs exhibit a flat and long potential plateau established due to the formation of stable chemical bonds between protons and oxygen in the oxide [17]. As a mechanism, the vacancies in proton-conductive perovskite-type oxides can be replaced by protons from oxygen ions to form substitution OH ion defects. The mechanism of hydrogen adsorption is physisorption, enhanced by an additional reaction mechanism due to the presence of redox species (Figure 5.15).

5.1.5.10 $Dy_2Sn_2O_7$

$Dy_2Sn_2O_7$ (dysprosium stannate), a binary oxide, crystallizes in the cubic Fd3m space group. The three-dimensional structure includes Dy trivalent cations bonded in a body-centered cubic geometry to eight O anions. Sn tetravalent cations are bonded to six equivalent O ions to form corner-sharing SnO_6 octahedra. Oxygens are bonded to four equivalent Dy ions to form ODy_4 tetrahedra that share corners with sixteen ODy_4 tetrahedra and edges with six equivalent ODy_2Sn_2 tetrahedra. Oxygens are also bonded to two equivalent Dy ions and two equivalent Sn ions to form a mixture of distorted corner- and edge-sharing ODy_2Sn_2 tetrahedra. Using chloride and nitrate pentahydrate precursors as Dy and Sn sources, the mixed solution can be calcined at ~500°C to create 20–25 nm porous NPs. The maximum discharge capacity of about 4000 mAh/g after 20 cycles indicates the mesoporous performance behavior of NPs [18]. The total reaction of the electrochemical hydrogen storage follows a reaction similar to that of other binary oxides as follows:

$$Dy_2Sn_2O_7 + xH_2O + xe^- \leftrightarrow \left(Dy_2Sn_2O_7 - xH\right) + xOH^-$$

The estimated 25 m^2/g SSA and 0.17 cm^3/g pore volume can be achieved in NPs with a combination of meso- and macropores [19]. The gradual increase in electrochemical

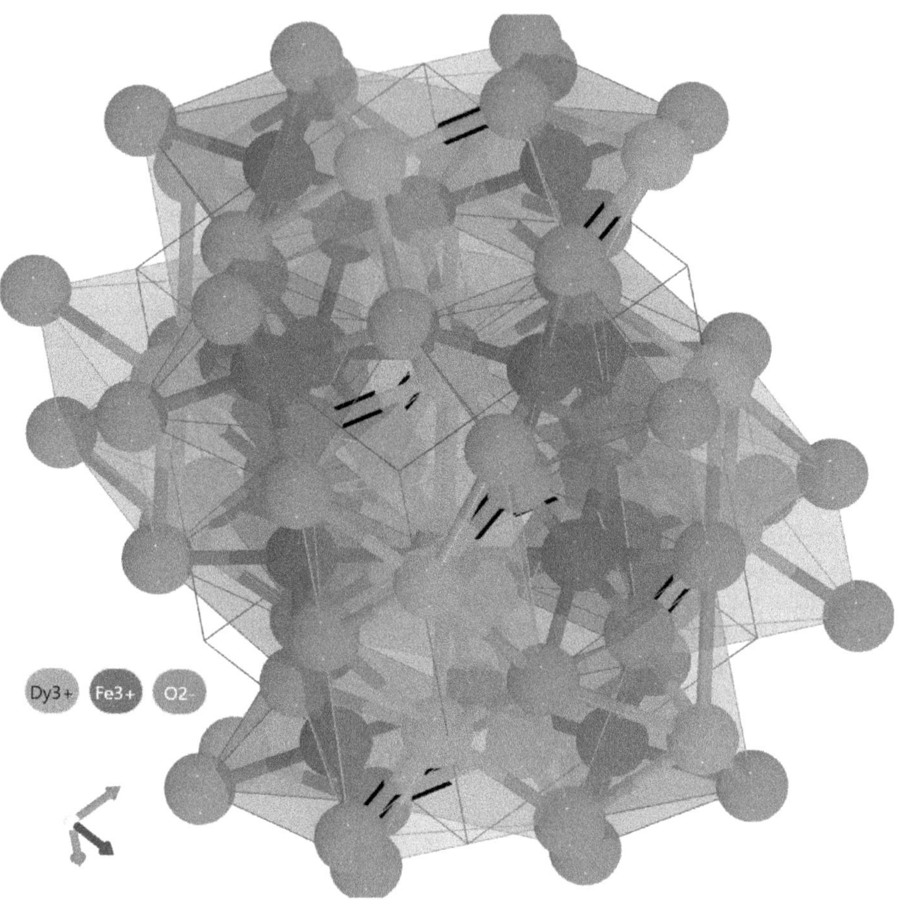

FIGURE 5.15 Crystal structure and ions positions of dysprosium–iron oxide. Created, designed, and introduced by the author.

hydrogen discharge can be related to the creation of new sites. The single plateau of the potential plot during charging indicates a one-step electrochemical charging by similar sites for H adsorption in the structure (Figure 5.16).

5.1.5.11　$Dy_3Al_5O_{12}$

The sol–gel auto-combustion synthesis using nitrate pentahydrate and aluminum nitrate nonahydrate with 800°C calcination leads to the formation of $Dy_3Al_5O_{12}$ NPs (nanogarnets) with a cubic Ia3d space group. The Dy trivalents are bonded in a distorted body-centered cubic geometry to eight equivalent O ions, with Al cations bonded to four equivalent O ions to form corner-sharing AlO_4 tetrahedra. The synthesized NPs are 45–55 nm in size and have a semi-spherical morphology. The electrochemical hydrogen SC of $Dy_3Al_5O_{12}$ can be estimated ~3140 mAh/g after 15 cycles following the below reaction:

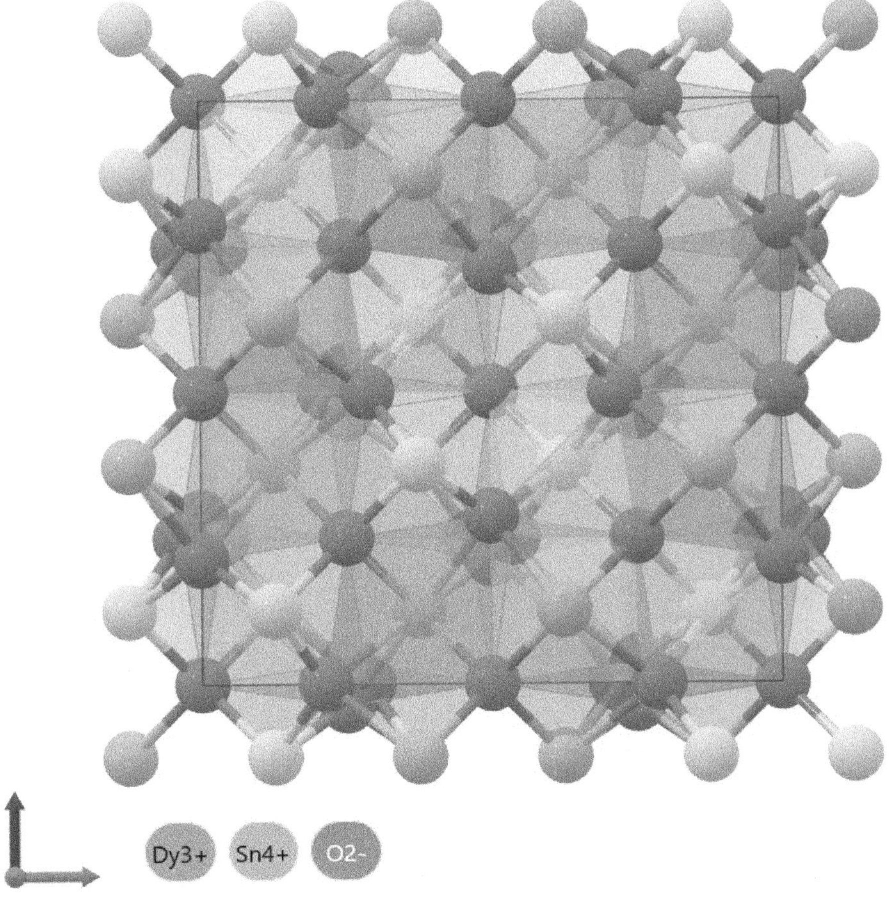

FIGURE 5.16 Crystal structure and ions positions of dysprosium stannate. Created, designed, and introduced by the author.

$$Dy_3Al_5O_{12} + xH_2O + xe^- \leftrightarrow (Dy_3Al_5O_{12} - xH) + xOH^-$$

as migrated hydrogen atoms adsorb to the surface. The presence of three plateaux of potentials on the discharging curve indicates three different electrochemical discharging steps and three different hydrogen adsorption sites. The stable chemical bonds between protons and oxygen in the oxide, along with the presence of vacancies in the structure of the MMOs, control the potential plateau, as the generated vacancies can be replaced by protons and oxygen ions to form substitutional OH ion defects (Figure 5.17).

5.1.5.12 $Sr_xCo_yO_z$

The strontium cobalt oxide, as a binary MMO, forms in a monoclinic Cm space group or an orthorhombic Pnma space group, with Sr divalent ions coordinated and bonded

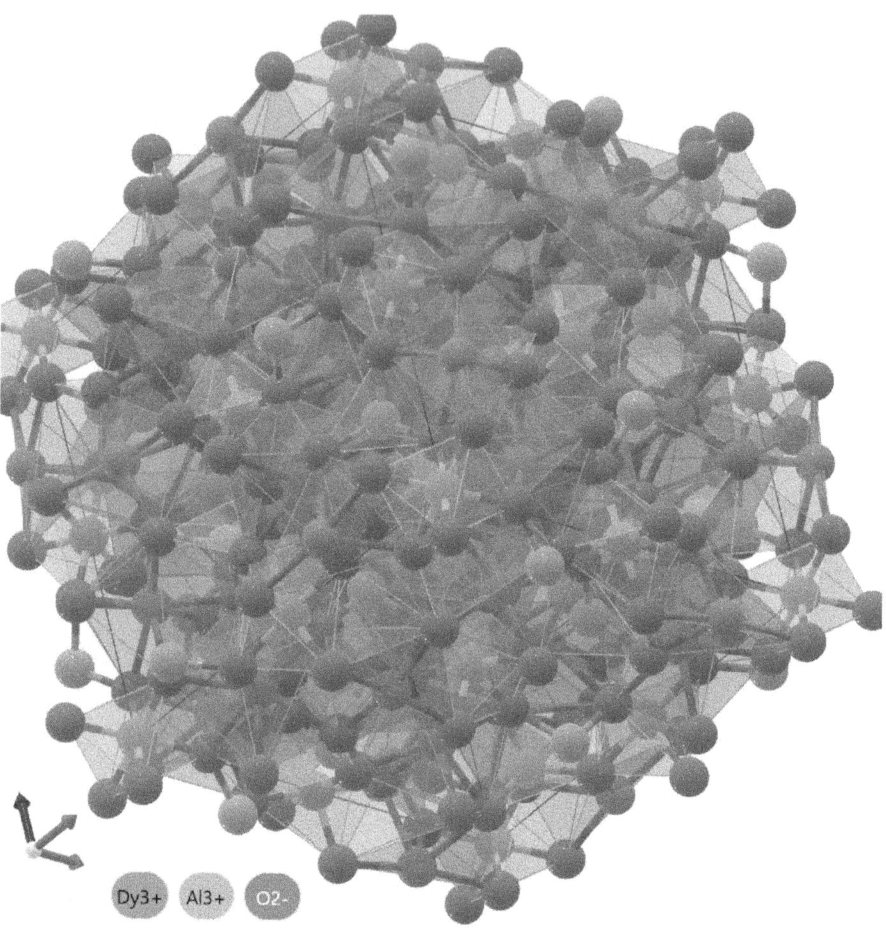

FIGURE 5.17 Crystal structure and ions positions of dysprosium aluminate. Created, designed, and introduced by the author.

in a nine-coordinate geometry to nine O anions. Cobalt trivalent ions are bonded to four O ions to CoO_4 tetrahedra that share corners with two equivalent CoO_6 octahedra and corners with two equivalent CoO_4 tetrahedra. Via auto-combustion synthesis from strontium nitrate and cobalt nitrate hexahydrate precursor solutions in the presence of organic fuel compounds (e.g., sucrose, lactose, and fructose), $Sr_xCo_yO_z$ NPs with nanometric sizes and a maximum discharge capacity of ~1000 mAh/g (long and flat potential plateaus) can be synthesized. The adsorption patterns match a type III isotherm with an H3 hysteresis loop, indicating aggregated particles with slit-shaped pores of non-uniform sizes and shapes. The total pore volume of 1.11 cm³/g, an average mean pore size of ~3 nm, and an SSA of up to 30 m²/g lead to hydrogen physisorption with a maximum discharge capacity of ~950 mAh/g after 15 cycles of discharging and a hydrogen content of ~3.5 wt.% through the following equation (Figure 5.18):

$$Sr_xCo_yO_z + xH_2O + xe^- \leftrightarrow \left(Sr_xCo_yO_z - xH\right) + xOH^-$$

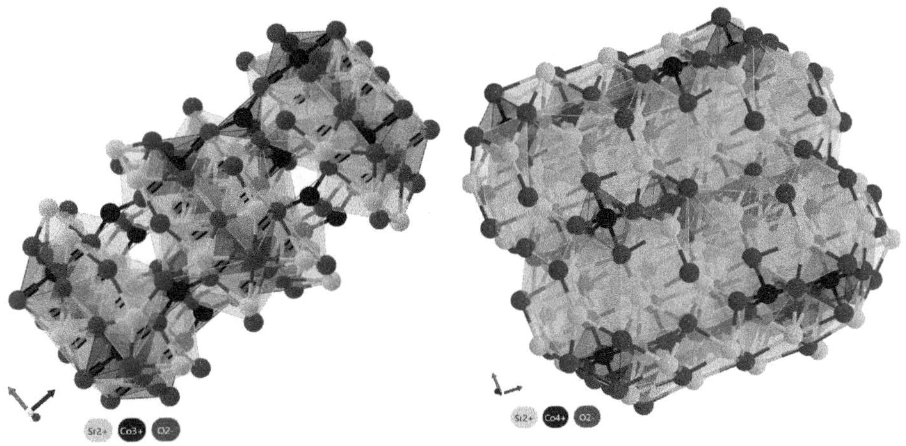

FIGURE 5.18 Crystal structure and ions positions of strontium cobalt oxide. Created, designed, and introduced by the author.

5.1.5.13 $Zn_2V_2O_7$

The zinc vanadate flower-shaped NPS can be synthesized through a sono-chemical method from zinc nitrate tetrahydrate and ammonium monovanadate precursor organic solution, followed by calcination at V^{5+} to V^{3+} and V^{4+}. The mesoporous NPs (type IV with H3-shaped hysteresis loops) create a SC of over 2900 mAh/g after 20 cycles, while three-dimensional spherical nanostructures show a SC of ~2300 mAh/g after the same number of cycles. The hysteresis loops are indicative of slit-like solids containing agglomerated particles [9]. The pore size distribution in the 10–40 nm range creates ~24 m²/g SSA and ~0.2 cm³/g pore volume. The electrochemical reaction of hydrogen adsorption is given as follows:

$$Zn_2V_2O_7 + xH_2O + xe^- \leftrightarrow (Zn_2V_2O_7 - xH) + xOH^-$$

with a charging step involving H creation through the decomposition of the alkali medium, the emigration of H atoms toward adsorption surfaces, and final adsorption [20]. The discharge stage involves the motion of H atoms from the electrode to the alkali medium. The discharge plateau potential of the NPs is estimated to be about 0.3 V, which is advantageous for electrochemical hydrogen storage (Figure 5.19).

5.1.5.14 $BaAl_2O_4/BaCO_3$

Similar to oxides, nanocomposite particles of oxides and non-oxides (i.e., $BaAl_2O_4/BaCO_3$) can be synthesized by solution combustion using nitrate precursors and organic maltose fuel, followed by high-temperature (700°C) calcination. The 20–30 nm NPs act as hydrogen storage systems with ~900 mA h/g discharge capacity and ~3.2 wt.% capacity, as follows:

$$BaAl_2O_4/BaCO_3 + xH_2O + xe^- \leftrightarrow (BaAl_2O_4/BaCO_3 - xH) + xOH^-$$

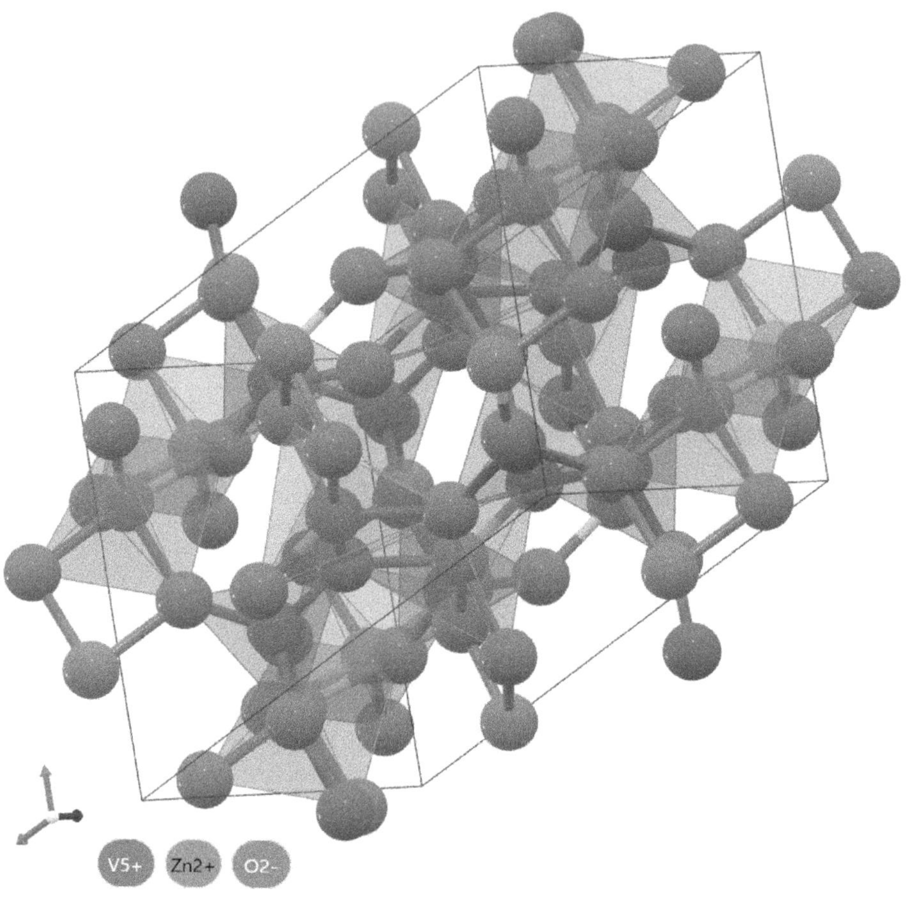

FIGURE 5.19   Crystal structure and ions positions of zinc pyrovanadate. Created, designed, and introduced by the author.

During the charging stage of the electrode, the electrolyte dissociates and the sample adsorbs hydrogen. The high hydrogen adsorption ability occurs as the polarization of the working electrode achieves larger values [21].

5.1.5.15   Ceria and CuO - CeO$_2$ Nanocomposites

Simple oxide systems such as ceria and copper oxide can be developed as hydrogen storage proton-conductive perovskite-like structures. As cerium can charge/discharge some hydrogen via the reduction characteristic of cerium (Ce^{4+}/Ce^{3+}), MMOs with other metals (M) can also be considered effective hydrogen adsorbents. The leading reaction is given as follows:

$$MCe_zO_y + xH_2O + xe^- \leftrightarrow \left(BMCe_zO_y - xH\right) + xOH^-$$

The composite NPs (15–30 nm) are synthesized through the thermal decomposition of copper and cerium nitrate solutions with 500°C calcination. The adsorption/desorption isotherm of NPs exhibits type IV behavior with H1-type hysteresis loops as significant features for ordered mesoporous materials. The hydrogen SC is related to a discharge capacity of ~2500 mAh/g (after 20 cycles), with a pore size of 5–6 nm, an SSA of ~85 m^2/g, and a pore volume of ~0.08 cm^3/g for a discharge capacity of ~1.5 wt.%. The plateau of discharge potential nearly 0.1 V is prolonged by the presence of CuO compared to pure ceria.

5.1.5.16 Co_2SnO_4

Cobalt stannate nanostructured hollow cubes (nanocubes) as a ternary transition metal oxide have conversion potential (~0.8 V) and theoretical capacity (1106 mA/g), which, alongside high SSA, makes them suitable candidates for hydrogen adsorption. Co_2SnO_4 a spinel-like structure that crystallizes in the orthorhombic Imma space group [22]. The Co divalent cations bond with four O atoms to form CoO_4 tetrahedra that share corner with six CoO_6 and SnO_6 octahedra [23]. They can be synthesized through sonochemistry using cobalt nitrite hexahydrate and tin chloride pentahydrate solutions, followed by 950 °C calcination and the addition of a black suspension of graphene to produce Co_2SnO_4-G nanocomposites. A slight increase in discharge capacities to 1190 mAh/g for Co_2SnO_4 NPs and more than 100% increase in Co_2SnO_4-G to ~2700 mAh/g after 25 cycles compared to theoretical levels can be achieved. The adsorption/desorption isotherms indicate a type IV isotherm (mesopores with a narrow pore size) with H2-type hysteresis, characteristic of solids with connected particles to cylindrical channels or formed by aggregates (consolidated) or agglomerates (unconsolidated) of spheroidal particles with non-uniform sizes or shapes [24]. The ~7 nm pore sizes are accompanied by ~34 m^2/g SSA and ~7.7 cm^3/g pore volume for ~3.3 wt.% discharge capacity (Figure 5.20).

5.1.5.17 Perovskites

Perovskite structures possess ionic orthorhombic crystal structures with a general formula of ABX_3 and are consist of metal cations (A and B) and a non-metal anion (X). The oxides, hydrides, and oxyhydrides are variations of perovskites used as hydrogen carriers (storage or compound). For instance, $MNiH_3$ perovskites (M: Li, Na, and K) have hydrogen capacities in the range of 3.5–4.5 wt.% with desorption temperatures in the 100°C–175°C range. The hydrogen SC of perovskites decreases with increasing M element mass. In $MNaH_3$ perovskite (M: Mn, Fe, and Co) systems, the negative formation energy implies their synthesizability and thermodynamic stability, with hydrogen storage capacities of 3.57, 3.70, and 3.74 wt.% for Mn, Fe, and Co, respectively, as X. The oxyhydride-type perovskites include TiO_3H_x, $CaTiO_3H_6$, and $MgTiO_3H_x$ with ~4.3 wt.% capacity at a nearly 550°C desorption temperature. Other systems such as $BaYO_3H_3$ show slightly unstable structures with a capacity of 1.09 wt.%.

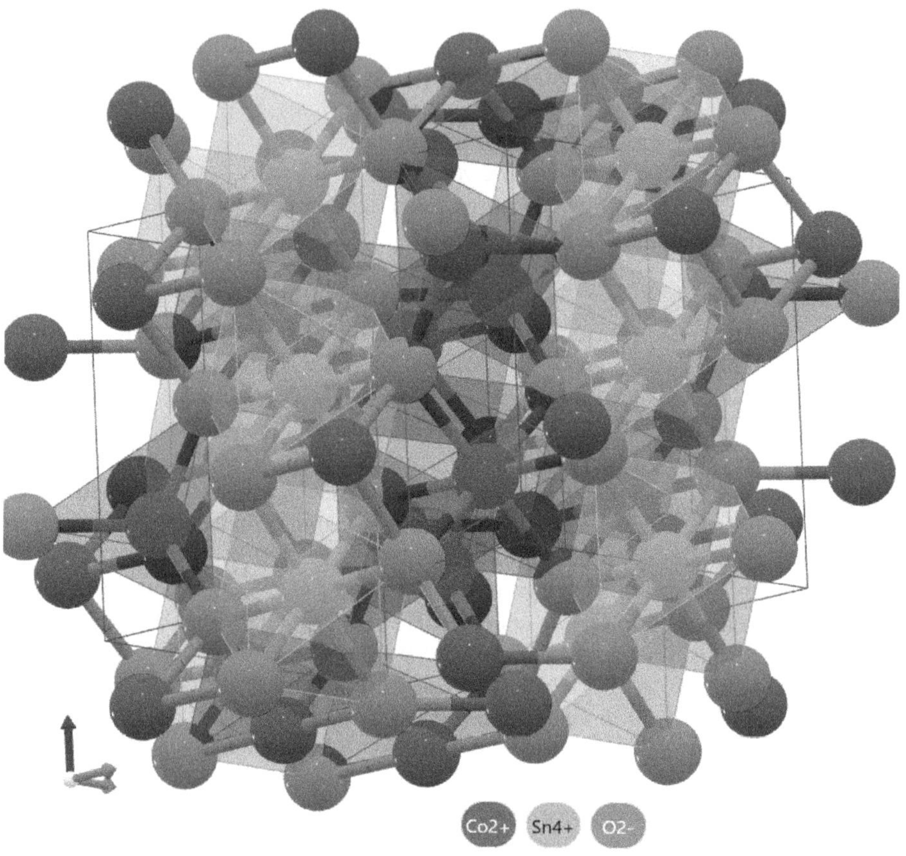

FIGURE 5.20 Crystal structure and ions positions of cobalt stannate. Created, designed, and introduced by the author.

5.1.6 Carbons

5.1.6.1 Activated Carbon and Carbon Aerogels

Activated carbon (AC), a low-cost material with distinguishable properties such as high SSA, high porosity, and desired surface functionalization, can be used for effective applications in adsorption. The basic chemical structure of AC closely approximates that of pure graphite (in its defective form). The graphite crystal is composed of layers of fused hexagons held together by weak vdW forces. The layers are connected by carbon–carbon bonds, with grains that have been treated to improve their absorptive ability. The structure includes a semi-crystalline arrangement with a hexagonal crystal structure and a pore structure of mesoporous size. Besides these advantages, the limitations include a lack of binding sites for hydrogen, non-uniform pore sizes, and difficulties in controlling pore structure [25]. Therefore, ACs consist of a twisted network of defective carbon layer planes cross-linked by aliphatic bridging groups. Carbon structures have a wide range of SSAs, from 2 to 4000 m^2/g, with narrow internal cavities or channels of microporous dimensions. Due to the heterogeneous structure of the surface features, especially for high-surface-area ACs, different types of binding sites are possible. The utilization of organic and/or inorganic compounds as modifiers facilitates the manufacturing of ACs with controlled porous structures and high yield factors. This developed advanced technology offers easy production of homogeneous activated carbon adsorbents with a pore volume of 0.3–0.6 cm^3/g (70%–80% volume of micropores), a SSA of up to 1500 m^2/g, and an adsorption capacity of up to 160 mg/g (3.5 MPa, 293 K) using benzene as a modifier. Unlike graphene, the surface areas of activated carbons can supply higher SSAs than graphite due to the zigzag and armchair terminations [26]. These terminations, as part of the structure of ACs, can contribute to the available surface area as hydrogen adsorption sites. Hydrogen and hydroxyl adsorption on these sites significantly decreases enthalpy. The mean heat of adsorption for activated carbons is estimated to be 7.8 kJ/mol H$_2$ over the first 0.8 wt.% hydrogen uptake. The impact of site heterogeneity and adsorbate–adsorbate interactions in activated carbons can be detected by the isoexcess isosteric heat of adsorption for a highly microporous activated carbon. The monotonic heat of adsorption decreases as a function of weight percentage uptake, which can be as high as about 10 kJ/mol and decrease to about 5 kJ/mol per hydrogen molecule. While microporous activated carbons have relatively high Henry's law enthalpies, the continuous decrease in enthalpy with adsorption creates complications due to possible hydrogen release under varying binding conditions. The armchair and zigzag structures at the edges of allotropes of carbon nanoparticles are shown schematically in Figure 5.21. The edge terminations, which contribute to the specific surface area, allow for values well beyond the ~2700 m^2/g of a single graphene sheet. The 1.2 cm^3/g pore volume is attributed to nearly 3000 m^2/g SSA with 60 mg/g hydrogen mass percent uptake; with a slope coefficient of 0.02 mass percent per SSA unit linear slope (1 mass percent uptake per 500 surface area units or 1 mass percent for every 0.2 of micropore volume unit) at −196°C/35 bar. The hydrogen SC is nearly 7 wt.% at ambient temperature and 500 bar pressure, which is the general hydrogen response attributed to AC. AC adsorbents can be manufactured through a two-stage

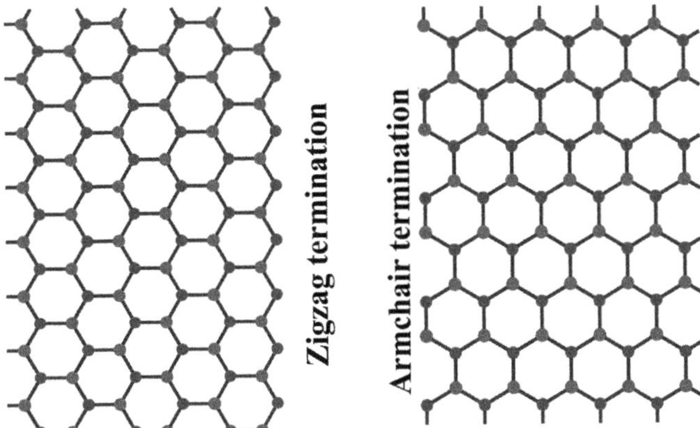

FIGURE 5.21 Armchair and zigzag structures at the edges. Created, designed, and introduced by the author.

activation process consisting of carbonization (with oxygen and hydrogen burning off) and gasification (heating charcoal in a steam or carbon dioxide atmosphere) to create a highly porous structure from carbon base precursor burn-off. The AC can be further activated (super-activated carbon) through the reaction of mesocarbon microbeads with alkaline hydroxides. The super-activated carbon has an open cage-like type of porosity and SSA as high as $3220\,m^2/g$. Reversible hydrogen adsorption following the characteristic Langmuir isotherm provides an SC of 5 wt.% H at $-196°C$ and 1.3 wt.% H at $25°C$. The complex AC based hydrogen adsorption systems include AC/TiC/MgH_2 with 5.6 wt.% at $300°C$/10 bar; AC/Mg/FV with 6.7 wt.% at $250°C$/10 bar; and AC/Li$(BH_4)_2$ with 1.0 wt.% at $400°C$/100 bar.

Carbon aerogels (CAs) are a category of amorphous carbons that share similar specifications with activated carbons. Their manufacturing involves the sol–gel polymerization process and the pyrolysis of an organic aerogel, consisting of nanosized particles interconnected to form a three-dimensional network, resulting in a material with high porosity and extremely low density [27]. CAs exhibit exceptional hydrogen adsorption properties due to large surface area, high porosity, and easy activation. Due to the more open structure of aerogels, the gravimetric densities are lower than those of activated carbons, mostly in 0.3–0.4 g/cm^3 range, with micropores smaller than 2 nm corresponding to $500–600\,m^2/g$ and nearly 5.5 wt.% gravimetric hydrogen uptakes at adsorption conditions of 77 K/30 bar. The complex CA-based hydrogen adsorption systems include CA/TiC/$NaAlH_4$, with 3.4 wt.% at $160°C$/90 bar; CA/$TiCl_3$/$NaAlH_4$ with 3.1 wt.% at $160°C$/60 bar; CA/Li$(BH_4)_2$ with 3.5 wt.% at $400°C$/100 bar; CA/Li$(BH_4)_2$ with 4.5 wt.% at $300°C$/100 bar; and CA/MgH_2 with 7.5 wt.% at $300°C$/60 bar [28].

5.1.6.2 Graphite and Graphene

Graphite, one of the most investigated allotropes of carbon, consists of alternating layers of sp^2-bonded trigonal planar sheets. These sheets are *graphene* sheets, with neighboring planes interacting through overlapping bonds between the unhybridized carbon $2p$ orbitals. The stacking sequence of the hexagonal graphite planes along the c-axis consists of an unsteady abab pattern, with half of the carbon atoms in any plane positioned between the hexagon centers of the layers above and below it. The carbon–carbon bond length in the basal plane is a =~1.42 Å, and the interlayer spacing is about 3.35 Å. Highly oriented pyrolytic graphite has a nearly perfect planar structure and is ordered with minimal interlaminar spacing. Graphite carbon allotropes may be nonporous, with specific surface areas of less than $20\,m^2/g$ and insignificant hydrogen uptake at low temperatures. Due to the dimensions of the graphite interplanar spacing and the size of molecular hydrogen, adsorption occurs only on the surface of graphitic carbons. The measured adsorption enthalpy of hydrogen on graphite is nearly 4 kJ/mol of hydrogen molecule. To increase hydrogen molecule uptake, the interplanar space (slit-pores) should be widened to accommodate guest molecules and decrease the enthalpy of adsorption, similar to designed graphene structures. As the potential fields from opposing slit-pore walls overlap, the heat of adsorption can be enhanced. The optimal interlayer spacing should be large enough to accommodate two hydrogen monolayers (i.e., one monolayer per slit-pore wall). Further interlayer expansion is not useful, as hydrogen only adsorbs in monolayers at supercritical temperatures. Although the graphene slit-pore structure is well suited for optimizing the carbon–hydrogen binding interaction in graphite, the gravimetric density remains intrinsically low due to the geometry. As SSA for each graphene layer can be estimated nearly 2700 m²/g for both sides of the layer, activating layers for adsorption drastically improves the gravimetric capacity in graphite structures. Since a single carbon hexagon contains a net total of two carbon atoms, the configuration of hydrogen on the graphite structure (graphene layer) includes a solid close-packed monolayer and a commensurate structure. Figure 5.22 provides a schematic depiction of the two configurations. The commensurate structures are energetically more favorable, as the hydrogen molecules form hexagons by positioning themselves at the centers. However, this configuration results in a lower gravimetric density than a close-packed structure. Through bidirectional adsorption on a graphene layer surface, the gravimetric density can be estimated at about 5 wt.% for the commensurate configuration. Meanwhile, the gravimetric density of a close-packed hydrogen monolayer can be estimated using theoretical cross-sectional area calculations for a hexagonal close-packed structure, where the cross-sectional area of a hydrogen molecule is calculated as follows:

$$\sigma = f\left(\frac{M}{\rho N_a}\right)^{2/3} \tag{5.1}$$

where 1.09 f factor for hexagonal close-packed structures, ρ is the density of liquid hydrogen, M is the molar mass, and N_a is Avogadro's number. Via the liquid hydrogen density of $\rho = 77.03$ kg/m³, the cross-sectional area is estimated about 0.124 nm². For the determined carbon SSA as c(SSA), hydrogen monolayer adsorption in weight percentage can be calculated as follows:

$$\text{wt.\%} = \left(\frac{c(\text{SSA})}{\sigma}\right)\cdot\left(\frac{M}{N_a}\right)\cdot 100 \tag{5.2}$$

which gives the amount of ~1.35 wt.% per 500 m²/g carbon surface area. Double-sided graphene has a surface area of c(SSA) about 2650 m²/g, indicating possible hydrogen adsorption of about 7 wt.%. These data show the theoretical limit for hydrogen

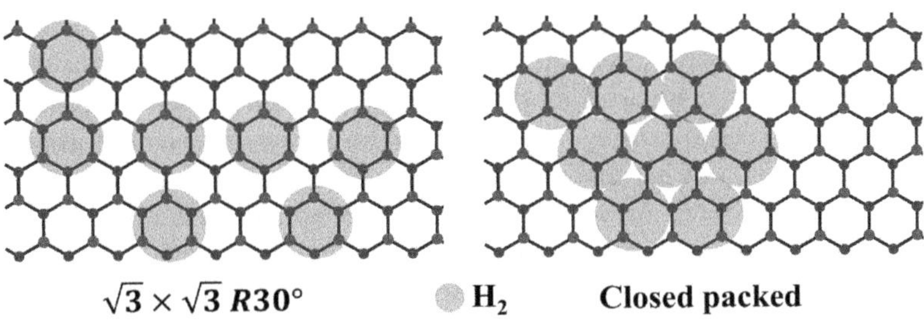

FIGURE 5.22 Schematical depiction of two configurations of hydrogen adsorption on the graphene layer or graphite outer surface. Created, designed, and introduced by the author.

density on graphene layers where both surfaces are accessible. Therefore, additional details must be considered to resolve the structure of hydrogen on graphite by using neutron surface studies to identify the actual molecular position of hydrogen with respect to the graphitic surface. Considering the volumetric densities of stacking molecular hydrogen in the c direction of a graphitic lattice, while the a-b stacking sequence is separated by 3.35 Å in graphite (the nearest-neighbor distance for molecular hydrogen is 3.51 Å), molecular hydrogen does not intercalate into graphite. As the bulk density of graphite is around 2.1 g/cm^3, the idealized slit-pore structure corresponds to the removal of two layers of graphite, resulting in an optimal slit-pore density of 0.7 g/cm^3 and a volumetric density of about 44 g/l. The synthesis of graphite with a slit-pore geometry requires pillars to separate the graphitic layers, which are typically held together only weakly via vdW attraction[29]. Further processing of graphite materials through ball milling and shearing of the layered structure at low temperatures of −200°C to −170°C and pressures up to 20 MPa results in 5.9–6.0 wt.% hydrogen adsorption. The hydrogen absorption in mechanically activated nanostructured graphite scales up to 7.4 wt.% H. These improvements are linked to changes in layer spacing due to stacking faults imposed by mechanical energy.

Based on graphene structure, two types of graphene can be considered: graphene with defects and graphene without defects. Polygraphene is also synthesized, consisting of two to nine layers of graphene sheets that are restricted along a plane; this is known as graphene nanoribbon. A multilayer combination of graphene nanoribbons is called graphene nanofibers. Graphene oxide (GO), as a chemically favored derivative of graphene, is obtained by the chemical oxidation of natural graphite or carbon nanofibers (CNFs) using strong oxidants. The basis for hydrogen adsorption via graphite and graphene structures lies in van der Waals interplanar and intermolecular interactions, as the p-orbital bonding between sheets does not hold graphene sheets together. The lobes of p-orbitals that belong to two consecutive sheets of carbon atoms do not overlap. The delocalized p-electrons repel each other and maintain a distance between the sheets. The sheets of carbon atoms are arranged parallel to each other and are held together by dominant vdW weak forces, which also keep graphene layers stacked. These attractions are similar to intermolecular forces in the physisorption mechanism of hydrogen molecules on carbon surfaces. vdW bonds are formed from intermolecular forces between atoms (molecules) and result from instantaneous charge distribution in atoms and molecules during approach. Under mutual interaction, an asymmetric polarization of electron charge is induced in molecules, creating temporary dipole moments, which attract atoms or molecules through electrostatic forces. These interactions occur between adjacent molecules and are insignificant compared to stronger chemical (covalent) bonds. The vdW interactions are essential as they determine bonding between layers of graphite and also the physisorption of hydrogen on carbon surfaces (in the kJ/mol range). In the absence of stronger bonds between Cs or C and H atoms, these atoms attract each other via quantum chemistry through dispersive interactions with a potential $V_{\text{LJ for C}}$ or $V_{\text{LJ for C–H}}$ proportional to r^{-6}. The chemisorption of hydrogen on graphite and graphene requires excess adsorption of hydrogen, above the limits set by the Langmuir isotherm and physisorption, with the formation of strong covalent bonds between C and H atoms. The highly defective structure of nanostructured graphite contains a large fraction of broken bonds, i.e., carbon dangling bonds, which can act as effective trapping sites for hydrogen storage. When the 1s-H orbital overlaps with an empty $2sp^2$-C orbital, a strong bond can be formed similar to an $2sp^3$-C orbital in diamondoids. Diamondoids are composed of cage hydrocarbon molecules that are superimposable on the diamond lattice. Chemisorption is most feasible on graphene plates in graphite structures. The primary chemisorption mechanism of hydrogen bonding can be observed in hydrofullerenes and hydrogenated carbon nanotubes (CNTs), with a peak in desorption above 400 K. By saturating the broken s bonds in graphene sheets and interacting with the p-electron density adjacent to the surface of the graphene plate, electrons from the 1s H orbital can form chemisorption. If p bonding were fully utilized, every carbon atom could serve as a site for chemisorption. Due to the strength of chemical s + p bonds, which are much stronger than vdW interactions, the adsorbed hydrogen can only be released at higher temperatures. Consequently, the chemisorption of hydrogen in graphitic nanocarbons is not ideal for reversible hydrogen storage. The complex graphite/graphene-based hydrogen adsorption systems include graphite/$NaAlH_4$ with 2.6 wt.% at 150°C/55 bar; graphite/$LiNi_5$ with 5.0 wt.% at 450°C/60 bar; graphene nanosheet/$NaAlH_4$ with 1.9 wt.% at 250°C/60 bar; graphene/$NaAlH_4$ with 5.6 wt.% at 180°C/80 bar; graphene nanofiber/$NaAlH_4$ with 3.6 wt.% at 150°C/40 bar; graphene/Mg/MgH_2 with 6.5 wt.% at 300°C/55 bar; graphene nanoplatelet/MgH_2 with 7.2 wt.% at 300°C/20 bar; graphene nanoplatelet/Fe_2O_3/MgH_2 with 6.2 wt.% at 290°C/15 bar; and graphite/TiFe with 0.75 wt.% at 210°C/20 bar.

5.1.6.3 Fullerenes

In order to enhance hydrogen binding in carbon adsorbents, curved carbon surfaces can be useful. Fullerenes, such as single-walled carbon nanotubes (SWCNTs) and C60 buckyballs, are formed from graphene-like sheets composed of five- or six-membered rings. The presence of pentagonal rings results in the curvature of the carbon planes, allowing the formation of C60 spheres. Similar to the hollow cylindrical structure of single- and multi-walled carbon nanotubes (MWCNTs) obtained by rolling up graphene sheets along different directions, fullerenes are formed by imposing curvature. As carbon nanotubes have similar hydrogen adsorption properties to activated carbons and other amorphous carbons, the only ordered allotropes of carbon that can adsorb hydrogen are fullerenes, which provide a unique opportunity as a nanocage with structurally efficient features for hydrogen storage. Fullerenes are classified based on their number of structural carbons as C70, C76, C80, and C84. See Figure 5.23 for some fullerene geometries. C60 is one of the most stable fullerenes and can be hydrogenated by several methods, such as the Birch reduction procedure, borane reduction, zinc–acid reduction, and catalytic hydrogenation. The highly hydrogenated fullerenes are not stable and desorb hydrogen after adsorption. Storing hydrogen inside a fullerene cage, as well as the interaction between the hydrogen and the storage media, creates a theoretical maximum hydrogen SC of 7.5 wt.% for C60. Large fullerenes (C60, C70, C78) and small fullerenes (C20, C24, C32) can be used for hydrogen adsorption. The complex fullerene-based hydrogen adsorption systems include C60/Li(BH$_4$)$_2$ with 4.2 wt.% at 350°C/12 bar; C60/Li(BH$_4$)$_2$ with 12.9 wt.% at 320°C/100 bar; and C60 with 5 wt.% at 350°C/100 bar.

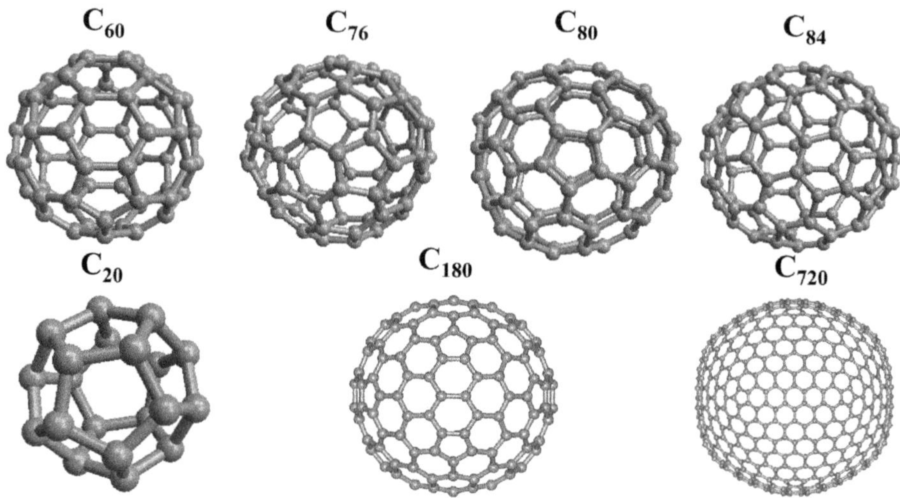

FIGURE 5.23   Schematical depiction of some fullerene geometries. Created, designed, and introduced by the author.

5.1.6.4   Carbon Nanotube

CNTs are nanocarriers that can be easily functionalized with certain chemical groups, as well as organic and inorganic materials. Functionalization of CNTs enhances their physical, chemical, and physiochemical properties. Due to their structural features, they serve as carriers for various molecules and atoms. The large surface area, physical dimensions, porosity, and various adsorbent/absorbent sites make CNTs suitable hosts for hydrogen storage. In both SWCNTs and MWCNTs, the arrangement of carbon atoms influences their configuration and morphology, such as zigzag, armchair, or chiral configurations in SWCNTs, and nanobuds, peapods, or cup-stacked structures in MWCNTs [30]. Hydrogen adsorption binding energies in CNTs are estimated based on defects. Since these defects are governed by configurations, larger adsorption energies can be achieved when the hydrogen molecular axis is perpendicular to the hexagonal carbon ring. The inorganic functionalization of CNTs using metals (M), MOs, and MMOs—such as the decoration of MWCNTs with rutile—improves hydrogen SC to about 2.0 wt.%. The formation of nanocomposites with $CoSe_2$/MOFs has enhanced hydrogen storage compared to individual components, as cobalt (Co) decorated CNTs achieve improved hydrogen storage capacities of 717.3 mAh/g (~2.6 wt.% hydrogen). Purified SWCNTs have a SSA of 1000 m²/g, with the hydrogen uptake at 77 K of just around 2 wt.%. The complex CNT-based hydrogen adsorption systems include MWCNT/LiNi₅ with 0.8 wt.% at 25°C/10 bar; TiC/CNT/MgH₂ with 7.0 wt.% at 300°C/10 bar; SWCNT/NaAlH₄ with 1.3 wt.% at 160°C/35 bar; SWCNT with 1.7 wt.% at 25°C/120 bar; MWCNT with 2.6 wt.% at 400°C/50 bar; and MWCNT/Ti₂Ni/Pd with 2.0 wt.% at 35°C/30 bar. Table 5.4 summarizes some of the typical CNT-based adsorption systems.

TABLE 5.4
Typical CNT-Based Adsorption Systems

System	SSA (m²/g)	Hydrogen capacity (wt.%)
Single-walled nanotubes	610–1000	0.55–1.1
Double-walled nanotubes	330–780	0.76–0.83
Multi-walled nanotubes	318–520	1.20–2.30

TABLE 5.5
Typical GNF-Based Adsorption Systems [31]

System	SSA (m²/g)	Hydrogen capacity (wt.%)
CNF	50–150	0.10–0.12
GNF	285–988	0.35–1.27
GNF–Herring bone	440–1050	1–1.18

5.1.6.5 Carbon Nanofibers, Whiskers, and Polyhedral Crystals

Graphitic polyhedral crystals are combinations of a high degree of order and small interplanar spacing without flat layers. In these structures, the graphite layers consistently wrap around a polyhedral core to form a crystal-like structure. Graphite fibers and whiskers have a tree-ring concentric cylinder morphology similar to the polyhedral graphitic particles, with a large aspect ratio and fibrous shape. They are primarily made of tubular graphene planes. The graphitic CNFs (GNFs) or CNFs consist of graphene planes arranged in platelet stacks in parallel or in angled arrangements that result in a conical fishbone structure. These characteristics indicate a large SSA and high hydrogen uptake values. For instance, a H:C ratio of up to 24 (6.8 wt.% H) in GNF herringbone fibers and 5.5 wt.% H in platelet stacks at 5°C/120 bar MPa is estimated. Table 5.5 summarizes some of the typical GNF-based adsorption systems.

5.1.6.5.1 Diamond and Nanodiamonds

In diamond, nanodiamonds, and diamondoids, the C–C bonding involves the valence electrons of carbon atoms filling the $2s$ and $2p$ orbitals, which represent electron distribution in 3D space as the square of the amplitude of their wave function. Since the valence electrons in the s- and p-orbitals are not equivalent, a nonsymmetrical and energetically disadvantaged molecule would result if the electrons filled their nonequivalent orbitals. Consequently, one of the 2s electrons is promoted to the 2p level, leading to four half-filled atomic orbitals. The s orbital combines with the p-orbital to create four equal sp3 hybridized orbitals, each with radially symmetrical s-type character and some directional p-type character. Each hybrid orbital contains one electron and overlaps with another carbon one-electron orbital, forming two-electron s-type bonds with other carbon atoms. The bonding of two electrons per bond results in the bond configuration of the diamond structure. This type of bonding is present in diamond and diamondoid nanophase carbons. The diamond structure features four nearest-neighbor carbon atoms bonded by s bonds. The three tetrahedral arms of the orbitals can be occupied by overlapping hydrogen atoms. Similar to a single carbon atom capped with tetrahedrally coordinated hydrogen atoms, a cluster of sp^3-bonded carbon atoms can be capped with hydrogen to form a hydrogenated diamond structure or diamondoids. Diamondoids can be considered cage-like saturated molecules that possess a rigid carbon framework of diamond-like sp3-bonded carbon atoms. Due to the rigid structure and high hydrogen content of diamondoids, they are viewed as potential building blocks for hydrogen storage. Despite the high hydrogen content in solid diamondoids, the storage mechanism involves chemisorption with the formation of strong bonds and lacks reversibility.

5.1.6.5.2 *Amorphous Carbon*

The nanocrystalline graphitic carbon can transform to amorphous carbon (am-C) if the size of sp^2-bonded clusters of carbon decrease below 1–2 nm. The am-C begins to exhibit any mixture of sp^2, sp^3, and even sp^1 sites. Amorphous carbon with a high fraction of sp^3 bonding like in diamond and diamondoids is known as diamond-like carbon. The amorphous carbon stabilizes its amorphous structure via hydrogen adsorption. Amorphous carbon thin films can contain as much as 40–60 at.% H. A lower hydrogen content of about 20 wt.% H results in an increased fraction of sp^2 hybridized bonds in the graphite-like hydrogenated amorphous carbon phase. Nearly 10 at.%. hydrogen can be released by breaking weak C–H bonds during low-temperature annealing (< 500°C) and 22 at.% by breaking covalent bonds at higher temperatures (~750°C).

5.2 CERAMIC CONFINEMENT SYSTEMS FOR HYDROGEN STORAGE

The concept of confinement is synonymous with partly or totally encapsulating or restricting materials in a small-sized environment (usually nanosized space, such as nanoconfined space) with pre-structured channels, two-dimensional interfaces, pores, pockets, and cavities at the nanoscale. Consequently, nanoconfined space plays an irreplaceable role in nanoconfinement. Terms such as *nanoconfinement* and *nano-encapsulation* are used to describe these systems, as well as *scaffolds*. Typically, a confinement system includes a host medium (bulk of the confiner), spaces (preformed voids or areas for adsorption and confinement), and guest materials (adsorbate or diffusing species). As shown in Figure 5.24, a combination of voids, guest species, and adsorbates of host nanoconfinement indicate the structure and function of confining systems for hydrogen adsorption.

Similar to nanostructured materials, nanoconfined spaces can be categorized into three types based on the dimensions of confinement: (i) longitudinal (one-dimensional), (ii) interfacial (two-dimensional), and (iii) porous (three-dimensional) nanoconfined spaces. Nanotubes are essential representatives of longitudinal confinement, featuring large internal and external surface areas as adsorption sites. MXenes, transition metal dichalcogenides, transition MOs, hydroxides, silicene, phosphorene, and graphene serve as examples of interfacial confinement,

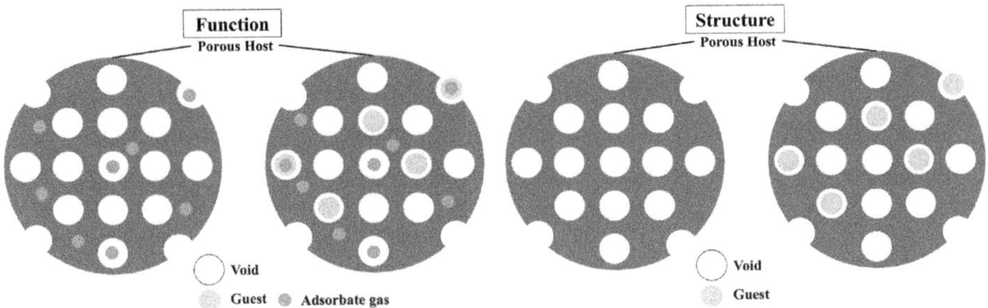

FIGURE 5.24 Concept of nanoconfinement for hydrogen adsorption. Created, designed, and introduced by the author.

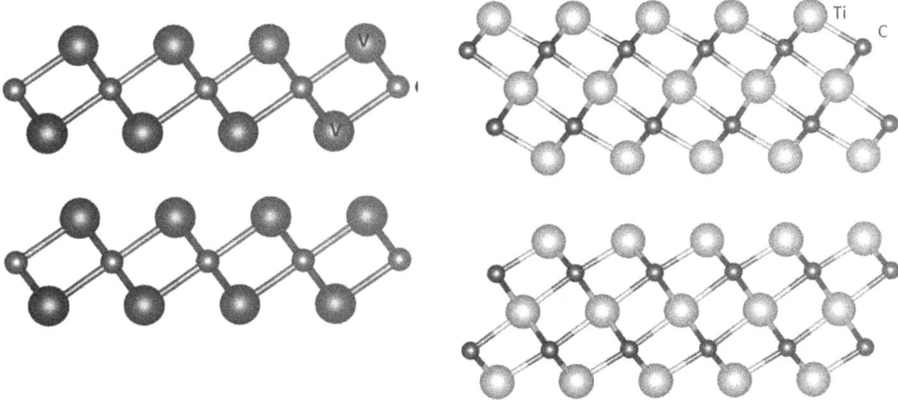

FIGURE 5.25 Schematic presentation of V_2C and Ti_2C MXene monolayers. Created, designed, and introduced by the author.

providing two-dimensional nanoconfined spaces between layers. The multilayered two-dimensional planes create large adsorption surface areas with accessibility through successive interplanar and over-planar routes, along with the potential for metal decorating or doping nanostructures [32]. Many theoretical studies on carbide and nitride MXenes have estimated impressive hydrogen storage properties (the highest among solid-state adsorption) that depend on composition and operational conditions. Through the interdiffusion of appropriate ions (mostly cations) in interlayer spaces, the size of the nanoconfined space can be adjusted (e.g., cation-π carbon interaction and the interaction between the hydrated cation and the oxidized groups in graphene). MXenes, as 2D materials composed of transition metal carbides, carbonitrides, and nitrides (with a general formula $M_{n+1}X_nT_X$ with M as a transition metal such as Sc, Ti, Zr, Hf, V, Nb, Ta, Cr, and Mo; X as C, N, P, and similar elements; and as surface termination molecules such as single-bond OH, single-bond O, or single-bond F groups) are also applicable hosts for confinement. A typical MXene sheet usually consists of 3, 5, or 7 atomic layers (M_2X, M_3X_2, or M_4X_3) [33]. Figure 5.25 shows the typical structure of V_2C and Ti_2C MXene monolayers. In the presence and bonding of TX groups to the outer transition metal layers,

TABLE 5.6

MXene Synthesis Summary Using Various Methods

Synthesis method	System	Example
Amine-assisted TBAOH method	Large-scale MXene	$Mo_2TiC_2T_x$
		$Cr_2TiC_2T_x$
		$Mo_2Ti_2C_3T_x$
Clay-based route (LiF/HCl etchants)	Clay-like MXenes	Clay-like $Ti_3C_2T_x$
Electrospinning method	Ti_3C_2	$Ti_3C_2T_x$
	$T_xMXene/carbon nanofibers$	MXene flakes
Intercalation/delamination method	Isolated single-layer MXene	Carbides
Mild method	Large flake MXenes	Single flake $Ti_3C_2T_x$
Oxygen-assisted molten fluoride salt method	Nano-layered Ti_2NT_x MXene	Ti_2AlN
Wet etching method	Various types of new MXenes	$(Ti,Nb)_2 CT_x$
		$(V,Cr)_3C_2T_x$
		$Ta_4C_3T_x$
Wet etching method (HF)	Ti_3C_2	$Ti_3C_2T_x$

there would be two more layers. Consequently, the interlayer space of layered materials and the interspace between closely packed crystal surfaces can be regarded as two-dimensional nanoconfinement space. Therefore, 2D materials have considerable potential due to their exceptional physicochemical characteristics compared to their bulk counterparts [34]. The three-dimensional spaces at the nanoscale and the related nanopores are examples of 3D nanoconfined space [35]. Some of the most common manufacturing methods are listed in Table 5.6.

5.3 MXENES FOR HYDROGEN STORAGE

The structures of layered 2D materials should be comparable to graphene, featuring a planar geometry and ultra-thin thickness. In a 2D flake of MXene, $n+1(n=1–3)$ layers of early transition metals are interleaved with n layers of carbon or nitrogen. With nearly 200 different MAX combinations used to manufacture MXenes and many new possible combinations from carbide and nitride mixed metal compounds, these 2D materials have real potential as ceramic hydrogen adsorption and storage systems [36]. The surface functionalities and the interlayer spacing of MXene sheets play a crucial role in hydrogen adsorption, as they influence the binding energy of hydrogen with MXene sheets. The initially adsorbed hydrogen during hydrogenation can split into H atoms through electron donation from 3d metal atoms to the hydrogen non-bonding orbital, followed by subsequent chemisorption phenomena. In contrast, further addition of hydrogen will retain its molecular form and get weakly adsorbed into the MXene sheet with a slightly elongated H-H bond distance. This combination of specifications is present in MXenes, allowing interaction with hydrogen through the primary adsorption mechanisms of physisorption (binding energy $<0.1\,eV$), chemisorption (binding energy $>0.8\,eV$), and the intermediate Kubas interaction (binding energy: $0.2–0.8\,eV$). The accessible Kubas binding interaction in MXenes improves their hydrogen SC during ambient condition adsorption and at the ambient-pressure/high-temperature (30–120°C) desorption stage, besides their lightweight and high SSA. The MXenes Kubas interaction occurs with electron donation from the σ bonding orbital of the hydrogen molecule to an empty d-orbital of the metal without disrupting the molecular bond. The unoccupied σ^* anti-bonding orbital of the hydrogen molecule accepts a π-back electron from the occupied transition metal d-orbital in the MXenes structure. The hydrogen SC can be distributed as ~40% physisorption, ~20% chemisorption, and ~40% Kubas adsorption. For instance, in carbide MXenes like vanadium carbide(V_2C), titanium carbide Ti_2C, and scandium carbide (Sc_2C) nanostructures; a maximum storage of nearly 8.5 wt.% with a similar distribution and ~0.12, 5.03, and 0.28 eV binding energy for physisorption, chemisorption, and Kubas interaction, respectively, has been recorded [37]. As the base MAX phases possess intrinsic potential for hydrogen storage, MXenes present numerous opportunities. The particular effects of MXene type, the variety of metals, and the interlayer spacing interactions, which are dependent on the surface and terminal functionalities, can indicate SC; they are also affected by procedural factors such as the extent of leaching and exfoliation, interlayer spacing, exposed surfaces, and the metal atoms forming MXene, as well as the incorporation of dopants. The synthesis of MXene hybrids, as a combination of hydrides and the MXene scaffold, can be a solution for improving hydrogen SC. Other parameters (such as MXene concentration, temperature, and pressure) should be further optimized to improve the de(hydrogenation) kinetics of MXene hybrids. Despite the improved initial hydrogen storage capacities, other limitations should be addressed, including (i) high hydrogen loss in high cycles (< 50 cycles), (ii) high onset desorption temperature, (iii) high desorption temperatures, (iv) low hydrogen absorption/desorption rates, and (v) small hydrogen sorption capacities [38]. Other minor issues are the lack of cycling data, limited information on isothermal hydrogen adsorption-desorption, and low reversibility. By incorporating mixed metal hydrides and alanates (guest phase) into different MXene structures (host phase), many improved hydrogen desorption/adsorption systems can be introduced [39]. The interactive forces and bonds between the metal hydrides/alanates and MXenes significantly increase hydrogen storage efficiencies (see Table 5.7). Studies on charge transfer and electronic density of states for hydrogen interactions with Ti_2N MXenes monolayer in a nine-supercell containing 9-N and 18-Ti atoms have shown the presence of Kubas, chemisorption, and physisorption of hydrogen (see Figure 5.26).

5.4 CERAMIC SCAFFOLD FOR HYDROGEN STORAGE

As hydrogen should be properly confined in porous structures for efficient adsorption, the gas-solid interface must be improved by increasing SSA and gas permeability channels and routes. Pre-designed and specially fabricated 3D structures of ceramic materials (scaffolds) can fulfill these requirements by carefully choosing the base material and fabrication process. The nanoconfinement in scaffolds includes nano-dimensional adsorbents or hydrides/alanates that have been confined within the nano-scaffold, generating active nano-interfaces that control the kinetic and thermodynamic limitations of metal hydrides/alanates.

TABLE 5.7

MXenes Hydrogen Storage Systems

MXenes type	Hydrogen capacity (wt.%)	Guest
V_2C	8.1	
Ti_2C	8.6	
Sc_2C	7.2	
Cr_2C	7.6	
Ti_3C_2	4.7	
MoS_2/Ti_3C_2-MXene@C	10.0	
Ti_2N	8.55	
$Ti_3C_2T_x$ multilayered 2D MXene	10.5	
ScYC	5.7	
Ti_2CT_x	8.8	
$K_2Ti_6O_{13}$	6.7	
$TiNbCT_x$	6.5	
Ti_2C	13.4	$LiBH_4$ $Mg(BH_4)_2$
Ti_3C_2	6.2	
$TiVO_{3.5}$	5.0	
Ti_2VC_2	10.0	
Ti_2VAlC_2	10.0	
Ti_3CN	7.0	
$K_2Ti_6O_{13}$	5.0	MgH_2
Ni-incorporated $Ti_3C_2T_x$	5.83	MgH_2
$TiO_2/Ti_3C_2T_x$	5.98	
$Nb_4C_3T_x$	6.01	
Cu@MXene	6.3	MgH_2
Ti_3C_2	6.9	AlH_3
V_2C	7.8	AlH_3
Ti_3C_2@Pt-3%	8.4	AlH_3
Ti_2C	10.2–13.2	$Mg(BH_4)_2$
Ti_3C_2	10.8	$Mg(BH_4)_2$
VF_4@Ti_3C_2	8.2	$Mg(BH_4)_2$
Ti_3C_2	5.37	$LiBH_4$
Ti_3C_2/NC	4.61	$NaAlH_4$
TiO_2/Ti_3C_2	4.97	$NaAlH_4$
Ti_2C	13.4	$LiBH_4$ $Mg(BH_4)_2$
Ti_3CN	8.8	$LiBH_4$ MgH_2
PrF_3/Ti_3C_2	7	MgH_2

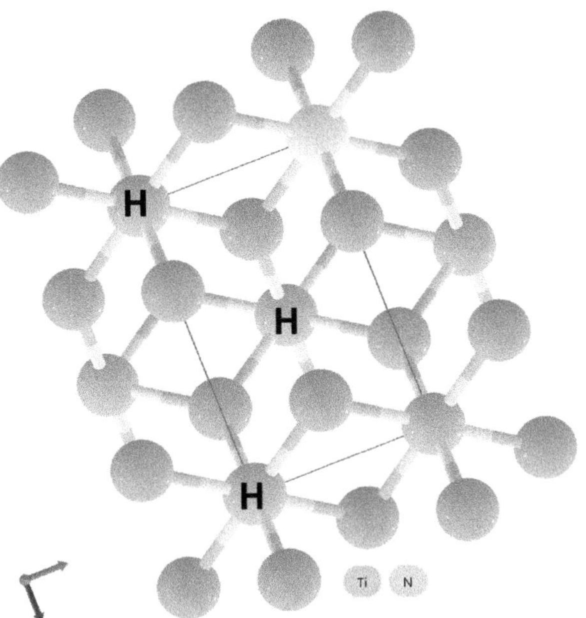

FIGURE 5.26 Schematic of single hydrogen molecule adsorption on monolayer MXene Ti$_2$N sites. Created, designed, and introduced by the author.

The concept of scaffolds can be expanded to ceramic foams. The specific target of scaffold functionality is achieved by adapting unique designs and shapes. The conventional manufacturing techniques include freeze-drying, solvent casting, electrospinning, sol–gel, replication foaming, and gas foaming. The replica technique utilizes a synthetic or natural template (e.g., polymeric foam) that is infiltrated with a ceramic suspension. After drying, the template is removed by heating, thus creating a replica of the original template structure. Many synthetic and natural cellular structures can be used as templates to fabricate macroporous ceramics through the replica technique. The *sacrificial template technique* incorporates a pore-forming agent or sacrificial material to act as a pattern former within the ceramic powder or slurry. Once the green body is formed, the part can be removed to leave empty pores. This method results in porous materials displaying a negative replica of the original sacrificial template, as opposed to the positive morphology obtained from the replica technique. In the direct foaming technique, gas bubbles are incorporated into a ceramic suspension, which, after slurry setting and drying, allows the ceramic to retain the resulting spherical pores. The dried objects, after pattern formation, should be sintered at high temperatures to obtain high-strength ceramic foams. The total porosity of the manufactured scaffolds is proportional to the amount of gas incorporated into the suspension or liquid medium during the foaming process or sacrificial template. The sizes of the pores may also depend on the stability of the wet foam before setting.

Recent progress in *additive manufacturing (AM)* has provided a platform to produce porous scaffolds with geometric freedom and the possibility of in situ development of functional structures. Numerous AM methods can be utilized to create tailored scaffolds, including L-PBF, material extrusion (ME), stereolithography (SLA), directed energy deposition (DED), binder jetting (3DP), and vat polymerization. Laser additive manufacturing (LAM) through the powder bed or L-PBF involves two techniques, namely SLS and SLM, collectively referred to as PBSLP. Nowadays, PBSLP has emerged as an attractive technique enabling various modes of functionalization of ceramic scaffolds [40].

The use of hydrides/alanates as materials for solid-state hydrogen storage can be enhanced by confining them within scaffold-like mesoporous ceramics, such as silica, graphene, and GO. The confined hydrides/alanates exhibit improved characteristics in hydrogen SC due to these confinements. Figure 5.27 schematically presents the effect of nano-scaffolding on the desorption capacity of typical hydride particles. For instance, silica is a sustainable and green material, capable of physical or chemical mixing with other compounds to form a highly active material for storing hydrogen. One of the promising types of nanoporous silica structures is amorphous silica [41]. Due to its significant surface area, mechanical stability, thermal stability, and chemical stability, it is a commonly used material for scaffolding. In comparison, the confined hydrides show enhanced dehydrogenation kinetics and desorption capacity, which improve with further regeneration cycles. The hydrogen capacity and kinetics of the hydrides can improve when appropriately confined in a porous structure. The confined hydride/alanate nanocomposites adsorb and desorb hydrogen at lower temperatures and in less time.

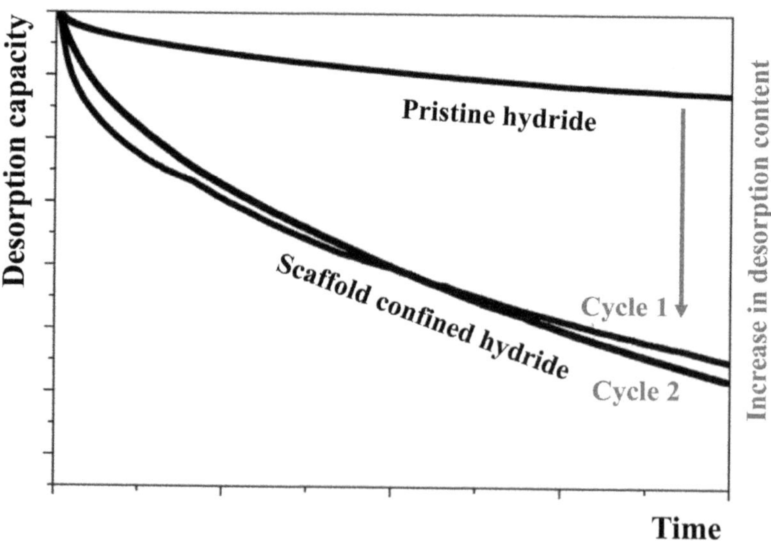

FIGURE 5.27 Schematic presentation of nano-scaffolding on the desorption capacity of a typical hydride particles. Created, designed, and introduced by the author.

5.5 CERAMIC ENCAPSULATION FOR HYDROGEN STORAGE

Another useful technique for improving hydrogen storage is the encapsulation of hydride/alanate material in a permeable porous medium, such as a porous capsule. Encapsulating hydrogen in a highly porous solid material results in a large dispersion of the hydrogen storage system. During encapsulation, hydrogen molecules or hydrides/alanates are encircled in the hollow centers of the host material, usually under high temperatures and pressures. The hydrogen molecules are contained within the voids of the host and are trapped under operating conditions until they are released due to a pressure drop or slight heating. The concept of multi- or single-walled encapsulation systems of single or multi-cored capsules is shown in Figure 5.28.

Furthermore, due to either excessively high or low decomposition temperatures, several metal hydrides/alanates do not exhibit acceptable electrochemical performance. By dispersing the storage materials inside small envelopes (encapsulation) and utilizing the highly porous nature of these materials, such as silica or other types of porous capsules, the kinetics of adsorption/desorption improves. This enhancement is related to minimized mass transfer distances. For instance, the encapsulation of lithium borohydride in carbon nanocages leads to high hydrogen uptake of the encapsulated hydrides after five consecutive cycles, with the encapsulated hydrides absorbing more hydrogen than non-encapsulated hydrides. Additionally, the infiltration of lithium borohydride into the porous channels of

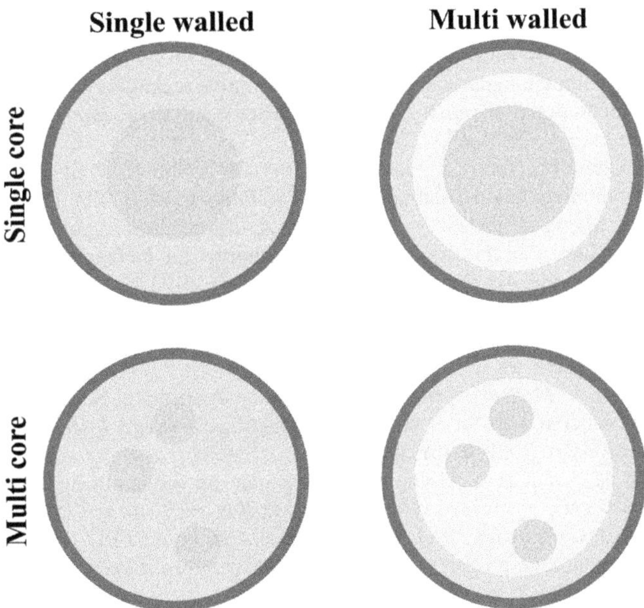

FIGURE 5.28 Schematic presentation of multi- or single-walled encapsulation systems of single- or multi-cored capsules. Created, designed, and introduced by the author.

carbon nanocages is another factor contributing to the good performance. In other systems, the synthetic conditions can regulate scaffold pore size, resulting in different-sized metal hydrides. Through the use of solvent or melt infiltration techniques or a solvothermal approach, metal hydrides can be integrated with scaffolds. The system variations and physicochemical properties of carbon scaffolds, such as ordered mesoporous carbons and graphene derivatives, allow for easy control of metal hydride nanoconfinement. For instance, sodium aluminum hydride can be confined in an ordered mesoporous carbon scaffold with over 80% high-capacity retention, as well as enhanced kinetics and cycling stability. Melt and wet infiltration are applicable methods used to introduce metal hydrides into a porous scaffold medium. For the hydride species to be adsorbed into the scaffold pores, the scaffold should be moistened either with a solution of hydride (high melting point hydrides) or with hydride in a molten state (low melting point hydrides), which uniformly distributes hydrides throughout the porous scaffold media. Furthermore, the presence of a scaffold improves the distribution of metal hydrides in the matrix pores and prevents the aggregation of nanoparticles. Some examples include zeolite-templated carbon encapsulation of NH_3BH_3 with an SSA of 3751 m^2/g and a capacity of 2 wt.%, SWCNT encapsulation of $NaBH_4$ with 2560 m^2/g SSA and 4.5 wt.% capacity, CNFs encapsulation of AlH_3 with 2213 m^2/g SSA and 6.5 wt.% capacity, MWCNT encapsulation of NH_3BH_3 with 2980 m^2/g SSA and 6.3 wt.% capacity, AC encapsulation of $LiBH_4$ with 3300 m^2/g SSA and 3.9 wt.% capacity, and CNT encapsulation of MgH_2 with 3076 m^2/g SSA and 0.5 wt.% capacity.

REFERENCES

1. Osman AI, Nasr M, Eltaweil AS, Hosny M, Farghali M, Al-Fatesh AS, et al. Advances in hydrogen storage materials: harnessing innovative technology, from machine learning to computational chemistry, for energy storage solutions. *International Journal of Hydrogen Energy*, 2024, 67: 1270–94.
2. Ng TYS, Chew TL, Yeong YF, Jawad ZA, Ho C-D. Zeolite RHO Synthesis Accelerated by Ultrasonic Irradiation Treatment. *Scientific Reports*, 2019, 9(1): 15062.
3. Sadeghi P, Myers M, Pareek V, Arami-Niya A. Temperature-regulated gas adsorption on LTA zeolites: Observation of sorption isotherms for methane, hydrogen, nitrogen and carbon dioxide. *Applied Surface Science*, 2024: 161037.
4. Timmerman MA, Xia R, Le PTP, Wang Y, ten Elshof JE. Metal oxide nanosheets as 2d building blocks for the design of novel materials. *Chemistry – A European Journal*, 2020, 26(42): 9084–98.
5. Zonarsaghar A, Mousavi-Kamazani M, Zinatloo-Ajabshir S. Sonochemical synthesis of CeVO4 nanoparticles for electrochemical hydrogen storage. *International Journal of Hydrogen Energy*, 2022, 47(8): 5403–17.
6. Tawalbeh M, Khan HA, Al-Othman A. Insights on the applications of metal oxide nanosheets in energy storage systems. *Journal of Energy Storage*, 2023, 60: 106656.
7. Wu H, Du J, Cai F, Xu F, Wei W, Guo J, et al. Catalytic EFFECTS OF V and V2O5 on hydrogen storage property of Mg17Al12 alloy. *International Journal of Hydrogen Energy*, 2018, 43(31): 14578–83.
8. Gholami T, Salavati-Niasari M, Sabet M, Abbasi A. Comparison of electrochemical hydrogen storage and Coulombic efficiency of ZnAl2O4 and ZnAl2O4-impregnated TiO2 synthesized using green method. *Journal of Cleaner Production*, 2018, 180: 587–94.

9. Masjedi-Arani M, Salavati-Niasari M. Novel synthesis of Zn2GeO4/graphene nanocomposite for enhanced electrochemical hydrogen storage performance. *International Journal of Hydrogen Energy*, 2017, 42(27): 17184–91.

10. Salehabadi A, Salavati-Niasari M, Sarrami F, Karton A. Sol-Gel auto-combustion synthesis and physicochemical properties of BaAl2O4 nanoparticles; electrochemical hydrogen storage performance and density functional theory. *Renewable Energy*, 2017, 114: 1419–26.

11. Rafiaei SM, Dini G, Bahrami A. Synthesis, crystal structure, optical and adsorption properties of BaAl2O4: Eu2+, Eu2+/ L3+ (L = Dy, Er, Sm, Gd, Nd, and Pr) phosphors. *Ceramics International*, 2020, 46(12): 20243–50.

12. Gholami T, Salavati-Niasari M, Varshoy S. Electrochemical hydrogen storage capacity and optical properties of NiAl2O4/NiO nanocomposite synthesized by green method. *International Journal of Hydrogen Energy*, 2017, 42(8): 5235–45.

13. Gholami T, Salavati-Niasari M, Salehabadi A, Amiri M, Shabani-Nooshabadi M, Rezaie M. Electrochemical hydrogen storage properties of NiAl2O4/NiO nanostructures using TiO2, SiO2 and graphene by auto-combustion method using green tea extract. *Renewable Energy*, 2018, 115: 199–207.

14. Gholami T, Salavati-Niasari M, Varshoy S. Investigation of the electrochemical hydrogen storage and photocatalytic properties of CoAl2O4 pigment: Green synthesis and characterization. *International Journal of Hydrogen Energy*, 2016, 41(22): 9418–26.

15. Barazandeh Behrooz T, Esmaeili-Zare M, Talkhab H, Behpour M. Ultrasonic-assisted synthesis of NiCo2O4/TiO2 ceramic as an efficient and novel hydrogen storage material. *Journal of the Iranian Chemical Society*, 2021, 18(10): 2613–23.

16. Salehabadi A, Salavati-Niasari M, Gholami T, Khoobi A. Dy3Fe5O12 and DyFeO3 nanostructures: Green and facial auto-combustion synthesis, characterization and comparative study on electrochemical hydrogen storage. *International Journal of Hydrogen Energy*, 2018, 43(20): 9713–21.

17. Salehabadi A, Salavati-Niasari M, Ghiyasiyan-Arani M. Self-assembly of hydrogen storage materials based multi-walled carbon nanotubes (MWCNTs) and Dy3Fe5O12 (DFO) nanoparticles. *Journal of Alloys and Compounds*, 2018, 745: 789–97.

18. Zinatloo-Ajabshir S, Morassaei MS, Salavati-Niasari M. Simple approach for the synthesis of Dy2Sn2O7 nanostructures as a hydrogen storage material from banana juice. *Journal of Cleaner Production*, 2019, 222: 103–10.

19. Zinatloo-Ajabshir S, Salehi Z, Amiri O, Salavati-Niasari M. Green synthesis, characterization and investigation of the electrochemical hydrogen storage properties of Dy2Ce2O7 nanostructures with fig extract. *International Journal of Hydrogen Energy*, 2019, 44(36): 20110–20.

20. Ashrafi S, Mousavi-Kamazani M, Zinatloo-Ajabshir S, Asghari A. Novel sonochemical synthesis of Zn2V2O7 nanostructures for electrochemical hydrogen storage. *International Journal of Hydrogen Energy*, 2020, 45(41): 21611–24.

21. Salehabadi A, Salavati-Niasari M, Gholami T. Effect of copper phthalocyanine (CuPc) on electrochemical hydrogen storage capacity of BaAl2O4/BaCO3 nanoparticles. *International Journal of Hydrogen Energy*, 2017, 42(22): 15308–18.

22. Zhang J, Liang J, Zhu Y, Wei D, Fan L, Qian Y. Synthesis of Co2SnO4 hollow cubes encapsulated in graphene as high capacity anode materials for lithium-ion batteries. *Journal of Materials Chemistry A*, 2014, 2(8): 2728–34.

23. Masjedi-Arani M, Salavati-Niasari M. Ultrasonic assisted synthesis of a nano-sized Co2SnO4/graphene: A potential material for electrochemical hydrogen storage application. *International Journal of Hydrogen Energy*, 2018, 43(9): 4381–92.

24. Erünal E, Ulusal F, Aslan MY, Güzel B, Üner D. Enhancement of hydrogen storage capacity of multi-walled carbon nanotubes with palladium doping prepared through supercritical CO_2 deposition method. *International Journal of Hydrogen Energy*, 2018, 43(23): 10755–64.

25. Slepičková Kasálková N, Slepička P, Švorčík V. Carbon nanostructures, nanolayers, and their composites. *Nanomaterials* (Basel, Switzerland), 2021, 11: 9.

26. Belhachemi M, Castro MMd, Casco M, Sepúlveda-Escribano A, Rodríguez-Reinoso F., editors. *Adsorption of hydrogen on activated carbons prepared by thermal activation: Hydrogen Storage, 10th International Renewable Energy Congress (IREC)*, 26–28 March, 2019.

27. Pang H-Q, Li S, Li Z-Y. Grand canonical Monte Carlo simulations of hydrogen adsorption in carbon aerogels. *International Journal of Hydrogen Energy*, 2021, 46(70): 34807–21.

28. Pandey AP, Bhatnagar A, Shukla V, Soni PK, Singh S, Verma SK, et al. Hydrogen storage properties of carbon aerogel synthesized by ambient pressure drying using new catalyst triethylamine. *International Journal of Hydrogen Energy*, 2020, 45(55): 30818–27.

29. Kopac T. Hydrogen storage characteristics of bio-based porous carbons of different origin: A comparative review. *International Journal of Energy Research*, 2021, 45(15): 20497–523.

30. Saxena S, Kumar Srivastava A, Riyazuddin SK, Samanta S, Khatua S, Singh A, et al. Single-walled carbon nanotubes patterned as aromatic element rings through chemical refluxation method. *Materials Today: Proceedings,* 2023. https://doi.org/10.1016/j.matpr.2023.01.300

31. Osman AI, Ayati A, Farrokhi M, Khadempir S, Rajabzadeh AR, Farghali M, et al. Innovations in hydrogen storage materials: Synthesis, applications, and prospects. *Journal of Energy Storage*, 2024, 95: 112376.

32. ur Rehman A, Akram Khan S, Mansha M, Iqbal S, Khan M, Mustansar Abbas S, et al. MXenes and MXene-based metal hydrides for solid-state hydrogen storage: a review. *Chemistry – An Asian Journal*, 2024, 19(16): e202400308.

33. Thomas T, Bontha S, Bishnoi A, Sharma P. MXene as a hydrogen storage material? A review from fundamentals to practical applications. *Journal of Energy Storage*, 2024, 88: 111493.

34. Saharan S, Ghanekar U, Meena S. V2N MXene for hydrogen storage: first-principles calculations. *The Journal of Physical Chemistry C*, 2024, 128(4): 1612–20.

35. Ahmad S, Mannan HA, Islam A, Nasir R, Qadir D. Nanostructured MXenes for Hydrogen Storage and Energy Applications. In Rizwan, K., Khan, A., Ahmed Asiri, A. M. (Eds.), *Handbook of Functionalized Nanostructured MXenes: Synthetic Strategies and Applications from Energy to Environment Sustainability*. Singapore: Springer Nature Singapore, 2023, pp. 155–71.

36. Zhou Z, Yu F, Ma J. Nanoconfinement engineering for enchanced adsorption of carbon materials, metal–organic frameworks, mesoporous silica, MXenes and porous organic polymers: a review. *Environmental Chemistry Letters*, 2022, 20(1): 563–95.

37. Shuaicheng HU, Honghui C, Xingbo HAN, Lijun LYU. Application of two-dimensional MXene in hydrogen storage. *Journal of Functional Materials*, 2022, 53(5): 5160–72.

38. Li Y, Guo Y, Chen W, Jiao Z, Ma S. Reversible hydrogen storage behaviors of Ti2N MXenes predicted by first-principles calculations. *Journal of Materials Science*, 2019, 54(1): 493–505.

39. Kumar P, Singh S, Hashmi SAR, Kim K-H. MXenes: Emerging 2D materials for hydrogen storage. *Nano Energy*, 2021, 85: 105989.

40. Abdulkadir BA, Jalil AA, Cheng CK, Setiabudi HD. Progress and advances in porous silica-based scaffolds for enhanced solid-state hydrogen storage: a systematic literature review. *Chemistry – An Asian Journal*, 2024, 19(2): e202300833.

41. Miri Z, Haugen HJ, Loca D, Rossi F, Perale G, Moghanian A, et al. Review on the strategies to improve the mechanical strength of highly porous bone bioceramic scaffolds. *Journal of the European Ceramic Society*, 2024, 44(1): 23–42.

6 Analysis and Assessment

Challenges, Opportunities, and Future

Scope

There are many techniques that have been introduced, developed, and perfected for evaluating the hydrogen adoption efficiency of systems. Methods include physical (morphology and size determination), chemical (bond and atom position), structural (powder diffraction), thermodynamics (thermal analysis and calorimetry), and kinetics (the time–temperature required for hydrogen absorption and desorption, and isotherms) assessments. The focus of this chapter is on methods that have typically been used for the characterization of hydrogen storage systems, with special emphasis on sorption measurement techniques and applicable methods for ceramics. Furthermore, the challenges, opportunities, and future of ceramic materials as hydrogen storage systems are discussed.

6.1 STORAGE ASSESSMENT

6.1.1 HYDROGEN GAS ADSORPTION ASSESSMENT

The measurement of gas adsorption is widely used for the characterization of powders and porous materials. It can be divided into two parts: (1) the constant volume method, in which the gas volume remains constant while the change in pressure is measured, and (2) the constant pressure method, in which the gas pressure remains constant while the change in volume is measured [1]. The determination of the mesopore size distribution is conventionally performed by methods based on the application of the Kelvin equation:

$$r_K = -2\gamma \upsilon_l / RT \ln\left(P/P_0\right) \tag{6.1}$$

which relates the Kelvin radius (r_K) to the relative pressure P/P_0 (at which vapor condenses in a cylindrical capillary) where γ is the surface tension and υ_l is the specific volume of the condensate [2]. The common *Brunauer–Emmett–Teller (BET)* method for surface area determination involves measuring nitrogen adsorption at a boiling point of 196°C, which indicates the specific surface area (SSA) of porous adsorbents. Related characterizations include the determination of the pore volume of porous solids through the measurement of nitrogen uptake, the determination of the pore size distribution (PSD) of the porous adsorbent, and the potential determination of porosity structures.

DOI: 10.1201/9781003488828-6

For *SSA determination*, the adsorption isotherm for a particular subcritical adsorptive is measured using volumetric apparatus, and the isotherm is analyzed based on a series of assumptions. The BET theory is an extension of the Langmuir theory of monolayer formation and accounts for the adsorption of additional multi-layers. The BET theory assumes adsorption on a flat surface with no limitation on the number of adsorbed layers, allowing for an infinite number of layers at $P/P_0 = 1$. It assumes that the surface is energetically homogeneous and that there is no interaction between neighboring adsorbate molecules. At equilibrium and after monolayer formation, the rate of condensation onto the first layer equals the rate of evaporation from the second layer. These adsorption and desorption rates depend on the equal energies of adsorption and liquefaction. The BET equation is given below:

$$V = \frac{V_m C P}{(P_0 - P)\left(1 + (C-1)\left(P/P_0\right)\right)} \tag{6.2}$$

where V is the adsorbed phase volume, V_m is the volume of a complete monolayer, C is the BET constant (+), P is the pressure, P_0 is the saturation pressure of the adsorptive, and P/P_0 is the relative pressure [3]. At low C values, the BET equation describes a Type III isotherm, while for higher values of C, it describes a Type II isotherm. The following equation from the BET equation determines the surface area of a material:

$$\frac{P}{V(P_0 - P)} = \frac{1}{V_m C} + \left(\frac{C-1}{V_m C}\right)\left(P/P_0\right) \tag{6.3}$$

By plotting $\dfrac{P}{V(P_0 - P)}$ against relative pressure P/P_0 (0.3–0.5) and determining the gradient m and intercept c of a linear curve, the gradient will be $m = \dfrac{C-1}{V_m C}$, the intercept $\dfrac{1}{V_m C}$, and

$$V_m = 1/(m+c) \tag{6.4}$$

and

$$C = \left(\frac{m}{c}\right) + 1 \tag{6.5}$$

to evaluate SSA with N_a as the Avogadro constant,

$$SSA = V_m N_a \sigma \tag{6.6}$$

For pore volume determination, adsorption by a porous material is due to a pore filling mechanism rather than Langmuir-type monolayer formation, and the adsorption capacity of an adsorbent depends on specific pore volume instead of SSA. The pore volume of a porous adsorbent is determined by measuring the adsorption isotherm.

If the isotherm exhibits Type I behavior, the pore volume can be determined from the saturation capacity.

For PSD, mesopore analysis is based on the Kelvin equation of the *Barrett–Joyner–Halenda* method. Other approaches applicable to microporous adsorbents include the *Dubinin–Radushkevich*, *Dubinin–Astakhov*, and *Horvath–Kawazoe* (HK) methods. The HK method describes the attractive solid–fluid forces in narrow pores but does not explain the mechanism of pore filling and does not account for fluid layering near the pore walls. Micropore filling is considered a continuous process, while the HK method assumes discontinuous pore filling at a specific pressure. Modern methods include *density functional theory* (DFT), and the *Monte Carlo* and *molecular dynamics* methods, which determine the equilibrium density profiles of confined fluids and subsequently calculate equilibrium adsorption isotherms from these densities[4]. The *grand canonical Monte Carlo (GCMC)* and *non-local DFT* methods are considered the most accurate methods for PSD determination. In the DFT method, local adsorbate density $\rho(r)$ at the position r is determined by minimizing the corresponding grand potential $\Omega(\rho(r))$. As the equilibrium form of $\rho(r)$ is determined, the isotherm and other properties of the system can be calculated. $\Omega(\rho(r))$ is given as follows:

$$\Omega(\rho(r)) = F(\rho(r)) - \int dr \rho(r)(\mu - V_{ext}(r)) \tag{6.7}$$

where F is the intrinsic Helmholtz free energy of the gas (the Helmholtz energy in the absence of an external fields), μ is the chemical potential, and $V_{ext}(r)$ is the potential field due to the pore walls. The description of gas–gas interaction allows for the reproduction of the bulk behavior of gas. The parameters that describe the solid–fluid interactions are obtained by fitting calculated isotherms to the standard isotherm on planar surfaces. The fluid is contained in individual pores with geometries such as slit-like or cylindrical. The DFT uses the local density approximation, which exempts short-range correlations, whereas non-local DFT employs a non-local smoothed density approximation that includes short-range correlations. The smoothed density approximation accounts for the oscillations of the density profile near the pore walls, providing an accurate non-local DFT description of gas behavior confined in narrow pores.

The GCMC method simulates the situation of an adsorbed gas in equilibrium with a bulk gas. GCMC is considered the most accurate method for describing confined matter in narrow pores. The GCMC and DFT methods produce isotherms calculated for a corresponding pore size within a specified range for the adsorbate under consideration. The isotherms are calculated by integrating $\rho(r)$ for the gas in pore sizes within the predefined range. The isotherm set (kernel) is used as the basis for the PSD determination. For $N(P/P_0, W)$ and P/P_0 as the relative pressure and W as the pore width (dimension), the generalized adsorption isotherm can be calculated as follows:

$$N\left(\frac{P}{P_0}, W\right) = \int_{W_{min}}^{W_{max}} N\left(\frac{P}{P_0}, W\right) f(W) \, dw \tag{6.8}$$

where $N(P/P_0)$ is the experimental isotherm for a relative pressure of P/P_0 and $f(W)$ is the PSD function. This equation assumes that the total isotherm consists of a number of individual isotherms for given pore sizes multiplied by their relative distribution. The accuracy of the PSD determination results depends on the availability of kernels that are applicable to the adsorbate–adsorbent system. The methods do not account for the chemical and geometric heterogeneity of pore walls. In hydrogen adsorbents, hydrogen can penetrate smaller pore sizes than argon or nitrogen. The combination of hydrogen and nitrogen adsorption is more advantageous for the PSD determination of microporous carbon allotropes, where hydrogen penetrates smaller pores that are inaccessible to nitrogen molecules, while nitrogen provides better sensitivity than hydrogen in mesoporous samples.

It is believed that the correlation between the BET surface area and hydrogen uptake is more relevant than the correlation between pore volume and hydrogen uptake. This assumption is reasonable, as any pores with large dimensions do not contribute significantly to the overall hydrogen adsorption capabilities of predominantly microporous materials. Therefore, the hydrogen storage capacity correlates reasonably well with micropore capacity, indicating that pore dimensions are critical to hydrogen capacity. The adsorbent with smaller pore dimensions adsorbs more hydrogen at lower pressure, whereas the material with a greater measured SSA adsorbs more hydrogen at saturation due to the greater total pore volume. At low pressures, hydrogen uptake correlates with the heat of adsorption; at intermediate pressures, it correlates with surface area; and at higher pressures, it correlates with the available pore volume. The pore volume must be considered in conjunction with pore size, as larger pore meso- or even macroporous hydrogen adsorbents contain higher pore volumes. Nevertheless, this does not lead to high hydrogen uptakes, as broader pores do not provide the enhanced adsorption potentials necessary for effective physisorption of hydrogen. Therefore, the BET method determines the micropore capacity rather than the surface area of a microporous hydrogen adsorbent for effective adsorption of hydrogen at elevated pressures.

6.1.2 THERMAL ANALYSIS AND CALORIMETRY

Thermal analysis and calorimetry are additional methods for hydrogen sorption measurements. Thermal analysis encompasses several methods, including thermogravimetric analysis and thermal desorption spectroscopy or temperature-programmed desorption (TPD). Other thermal analysis techniques include differential thermal analysis (DTA) and differential scanning calorimetry (DSC), along with reaction calorimetry. Thermogravimetric analysis is a method of thermal analysis in which changes in the physical and chemical properties of materials are measured as a function of increasing temperature (at a constant heating rate) or as a function of time (at a constant temperature and/or constant mass loss). It precisely measures the thermal stability and decomposition behavior of materials, as well as the decomposition rate during the heating and cooling cycles of the material. Thermal desorption spectroscopy is used to study hydrogen permeation in materials and involves controlled heating to measure the amount of diffusible hydrogen released from the material. TPD is used to monitor events occurring at the surface of a solid (adsorption/

desorption), while the temperature of the sample is changed according to a programmed schedule. It involves the characterization of energetic properties of adsorbents and studies surface interactions between molecular species on a surface when the surface temperature changes in a controlled setting. DTA uses the temperature difference between the adsorbent and a reference material subjected to the same heating program. During thermal cycles (having the same heating or cooling program), it records any temperature difference between them. DSC measures temperatures and heat flows (enthalpy) associated with thermal transitions during adsorption. Heat flux DSC is similar to DTA but determines heat flow between the sample and the reference material instead of temperature difference. In power-compensated DSC, the sample and the reference material are heated independently but at the same temperature ramp rate. The phase transition or reaction in the sample results in a peak in heat flow as a function of time. In hydride phase formation or decomposition, the recorded signal can indicate the enthalpy of hydride formation or decomposition. The temperature at which a DSC change is recorded provides the temperature of desorption. The heat–time–temperature plots contain information about the PSD of adsorbents, as schematically shown in Figure 6.1.

Temperature and enthalpy of desorption (the activation energy of the hydrogen desorption) can be determined using temperature-related methods such as *Kissinger* plots according to the following equation:

$$\ln\left(\frac{\beta}{T_p^2}\right) = \left(-\frac{E_a}{R}\right)\frac{1}{T_p} + \ln\left(\frac{AR}{E_a}\right) \tag{6.9}$$

where β is the heating rate, T_p is the peak temperature obtained from TPD, E_a is the activation energy, A is a constant, and R is the universal gas constant. The activation

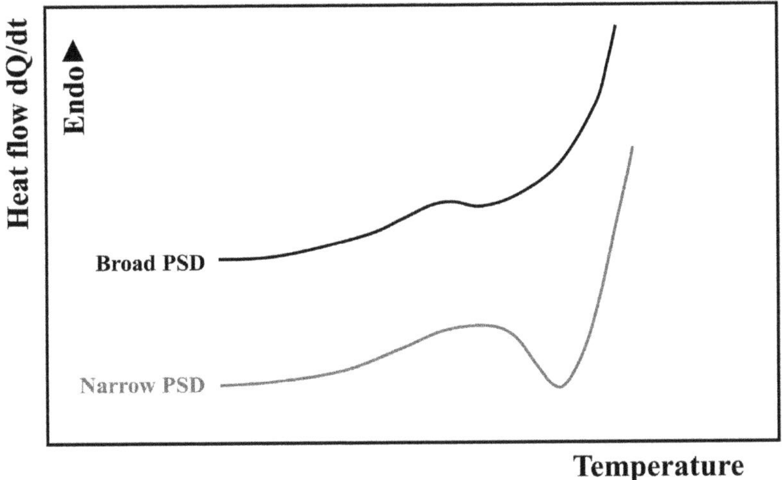

FIGURE 6.1 DSC plots for samples with narrow and broad pore size distributions. Created, designed, and introduced by the author.

energy is usually estimated from linear $\ln\left(\dfrac{\beta}{T_p^2}\right)$ versus $\dfrac{1}{T_p}$ as the slope of the plot [5].

Reaction calorimetry is used to measure the amount of energy consumed or released during a chemisorption reaction, which can be implemented in several ways, including heat flux and power-compensated variants. It provides higher accuracy than DSC. Using calorimetry, the determination of the enthalpies of hydrogen solution can be carried out via pressure–composition isotherm measurement. By combining calorimetry with volumetric sorption, detailed thermodynamic studies as a function of hydrogen content can be performed. Other thermal analysis methods can be hyphenated into hybrid techniques with additional methods. These combinations (DSC and MS signals) create thermal and adsorption/desorption match plots, as shown schematically in Figure 6.2. The combination of Langmuir isotherms with thermal analysis (Van't Hoff relation mentioned in chapter three) leads to heat-dependent surface coverage θ as follows:

$$\theta = \frac{P\exp\left(\dfrac{\Delta S^0}{R}\right)\exp\left(-\dfrac{\Delta H^0}{RT}\right)}{1 + P\exp\left(\dfrac{\Delta S^0}{R}\right)\exp\left(-\dfrac{\Delta H^0}{RT}\right)} \tag{6.10}$$

where P is the pressure; ΔS^0 and ΔH^0 are the entropy and enthalpy changes relative to standard pressure, respectively; R is the universal gas constant; and T is the temperature [6]. In the linear form, it calculates θ as follows:

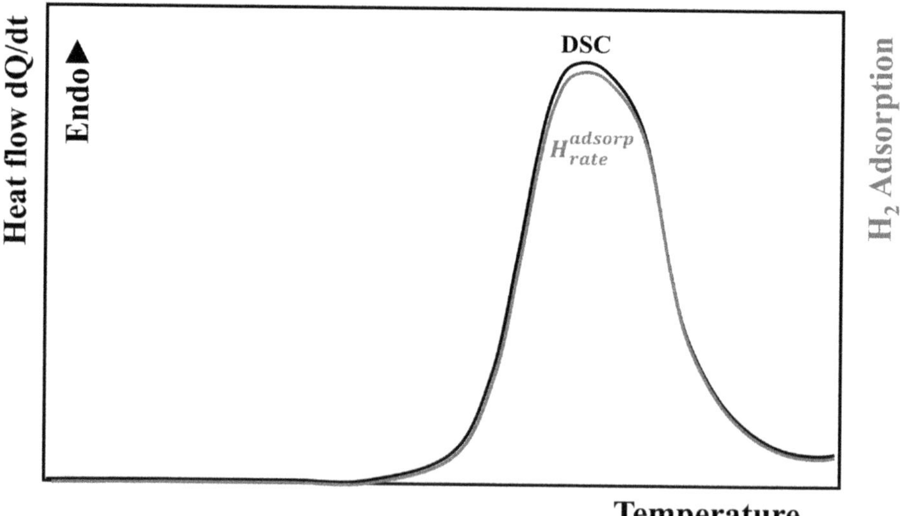

FIGURE 6.2 Schematic DSC plots accompanied by hydrogen desorption rates show significant match. Created, designed, and introduced by the author.

$$\ln\left(\frac{\theta}{P(1-\theta)}\right) = -\frac{\Delta H^0}{RT} + \frac{\Delta S^0}{R} \tag{6.11}$$

Thermoporometry is used to study the thermal interaction between an adsorbent and gases and provides information about its porosity and surface area. Also known as thermoporosimetry, it uses a calorimetric method that determines pore size (r_P) based on the melting or crystallization point depression (ΔT) of a liquid confined in a pore according to the following equation:

$$r_P = -64.67 / \Delta T + 0.57 \tag{6.12}$$

and melting energy (heat of melting) as follows:

$$W_a = -0.155\Delta T^2 - 11.39\Delta T - 332 \tag{6.13}$$

It can be used to indicate total porosity (percentage of empty space to the total volume), connected porosity (linked void spaces), disconnected porosity (isolated void spaces), effective porosity (fraction of interconnected void spaces), trapped porosity (void spaces inaccessible or sealed off from fluid flow), and dissolution porosity (porosity resulting from the removal of certain compounds) [7].

Temperature-programmed reduction is used to measure hydrogen interaction with surfaces, specifically surface reduction. In the presence of hydrogen inert substrates (non-hydrogen reactive surfaces), the hydrogen consumption during a controlled heat treatment regime can be used to assess hydrogen adsorption capacity and possibly hydrogen inter-diffusivity by acquiring quantitative and qualitative data from mass spectrometry. Analysis usually includes high-temperature degassing of adsorbents under an Ar atmosphere, adsorption under hydrogen pressure at different temperatures, cooling under hydrogen to extra low temperatures (~−100°C), and purging with Ar. The hydrogen desorbed upon heating to room temperature is used to estimate hydrogen capacity and kinetics (time-dependent). The schematic temperature and time dependency of hydrogen capacity in a typical hydrogen adsorbent is shown in Figure 6.3. As seen, adsorption at higher temperatures has resulted in less room temperature desorption, indicative of higher hydrogen adsorption capacity.

6.1.3 SPECTROSCOPY

The different methods of hydrogen storage spectroscopy measure and interpret electromagnetic spectra collected from adsorbents after stimulation. Techniques may include X-ray spectroscopy, atomic emission spectroscopy, atomic absorption spectroscopy, visible and ultraviolet (UV) spectroscopy, infrared (IR) and near-IR (NIR) spectroscopy, nuclear magnetic resonance (NMR), inelastic neutron scattering (INS), quasi-elastic neutron scattering (QENS), and Raman spectroscopy.

X-ray spectroscopy is used as a spectroscopic technique involving X-ray excitation to study molecules. The applicable subcategories of X-ray spectroscopy for hydrogen storage analysis include soft X-ray techniques such as X-ray absorption

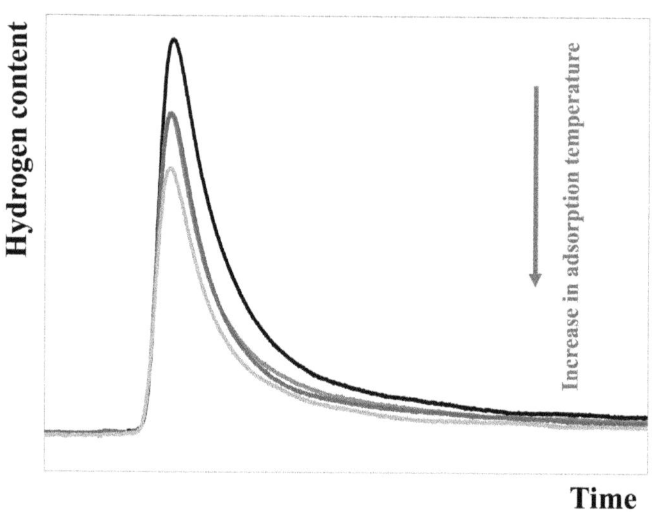

FIGURE 6.3 Schematic of hydrogen adsorption/desorption response via temperature-programmed reduction. Created, designed, and introduced by the author.

spectroscopy, scanning transmission X-ray microscopy, and ambient pressure X-ray photoelectron spectroscopy.

Atomic absorption spectroscopy is used to indicate elements in either liquid or solid samples via electrothermal vaporization and through the application of characteristic wavelengths of electromagnetic radiation. Similarly, *atomic emission spectroscopy* quantifies elemental presence through the intensity of light emitted from a flame, plasma, arc, or spark at a particular wavelength. These emissions' colors and wavelengths include red (656.2 nm), blue-green (486.1 nm), blue-violet (434.0 nm), and violet (410.1 nm).

In *visible and UV spectroscopy*, a type of absorption spectroscopy is utilized in which UV–visible light is absorbed by molecules. Absorption of UV–visible radiation results in the excitation of electrons from lower to higher energy levels, which, in the case of hydrogen bonding, shifts the UV absorptions toward shorter .

IR and NIR spectroscopy use the IR (780 nm to 1 mm) and NIR regions of the electromagnetic spectrum (from 780 nm to 2500 nm) to measure the interaction of IR radiation with matter by absorption, emission, or reflection, and to study and identify chemical substances or functional groups in solid, liquid, or gaseous forms. Since homonuclear diatomic molecules such as hydrogen do not have a permanent dipole moment nor do they exhibit stretching of atoms at the bond, they do not show any IR spectra separately. IR spectroscopy can detect heterolytically adsorbed hydrogen molecules at the adsorbent substrate. This technique can be used to determine the energy of gas–solid interactions, as the adsorptive molecule has at least one IR absorption mode that appears as a result of its interaction with the surface during adsorption [8].

NMR is a method for the determination of diffusion coefficients and jump rates for hydrogen in adsorbents. This spectroscopy can use any isotope with a nuclear magnetic moment and can therefore be performed with hydrogen, deuterium, and tritium. It has been used for the study of microstructural composition and evolution of complex hydrides during or as a result of hydrogenation and dehydrogenation. In NMR spectroscopy, a radio frequency magnetic field is applied in the presence of an applied static field, while the response is detected by a receiver. The technique can involve the measurement of the position or frequency of the resonance line, the strength of the line, the linewidth, or one of the relaxation times. The study of hydrogen interaction involves the measurement of relaxation times, pulsed field gradient methods, and the determination of linewidth.

INS is used to study the dynamics of hydrogen in adsorbents and the adsorption of molecular hydrogen in microporous adsorbents. INS involves the determination of the energy transfer between neutrons and the scattering objects. This technique can be used to study atomic hydrogen or to determine the quantized excitation energies of adsorbed molecular hydrogen. The scattering can be coherent or incoherent. The penetrating nature and strong scattering by hydrogen relative to other heavier elements are the main advantages of neutron scattering techniques. INS has the additional advantage of being independent of long-range crystalline order in samples. In microporous adsorbent characterization, INS may be used to observe the rotational transitions of adsorbed molecular hydrogen and to identify five different adsorption sites from the observed rotational transition peaks. INS spectroscopy can be combined with first principles calculations to aid in the assignment of peaks in the neutron energy transfer spectra to particular transitions. For instance, stabilized clathrate hydrates can be studied by INS to utilize its ability to observe the behavior of hydrogen confined in the cages.

QENS, as a limiting case of INS, is an INS-related technique in which the broadening of the elastic scattering peak is measured to determine information regarding the diffusion rate of hydrogen into the substrate.

Raman spectroscopy is an analytical technique in which scattered light is used to measure the vibrational energy modes of adsorbents. When light is scattered by a molecule, the oscillating electromagnetic field of a photon induces a polarization of the molecular electron cloud, leaving the molecule in a higher energy state with the energy of the photon transferred to it. In this method, hydrogen diffusion results in Lorentzian broadening of the elastic scattering peak, and the overall peak shape can be represented by one or more Lorentzian functions. QENS can be used to study hydrogen dynamics in microporous adsorbents such as carbon allotropes, zeolites, MOFs, microporous amorphous silica, and hydrides, such as alanates [9]. This technique allows for direct observation of hydrogen in a particular sample and the determination of hydrogen diffusion coefficients, which are related to the kinetics of the hydrogen sorption process, without quantifying hydrogen content. The variable temperature IR spectroscopy has been developed to determine the enthalpy of adsorption in microporous adsorbents, as the integrated intensity of the corresponding absorption band is proportional to the surface coverage. This can be achieved (i) by recording the integrated intensity of the IR absorption band and the equilibrium hydrogen pressure at variable temperatures using a Fourier transform IR spectrometer [10] and (ii) by employing a pseudo-isobaric approach to record the integrated intensity of the IR absorption band at variable temperatures.

6.1.4   DIFFRACTION

These methods for hydrogen storage characterization use the adsorbent's interaction (diffraction and elastic scattering instead of adsorption) with electromagnetic radiation (X-ray, neutron) such as X-ray diffraction and neutron diffraction.

In both X-ray and neutron diffraction, the incoming beam is elastically scattered, producing peaks at certain angles controlled by the radiation wavelength, the scattering angle, and the lattice spacing for a particular set of crystal planes. The X-ray interaction is governed by the scattering strength of the electrons in the scattering atoms. The strength of the scattering (scattering length or scattering cross section) depends on the number of electrons surrounding the nucleus; consequently, the larger the atom, the greater the X-ray scattering strength. X-ray scattering by light atoms, such as hydrogen, can be overwhelmed by the scattering strength of heavier surrounding atoms. Neutron scattering is governed by the interaction strength between the incoming neutrons and the nucleus of the scattering atom. As a result, the scattering strength by atomic mass varies in an unpredictable manner, and the scattering strengths of elements close in atomic mass can therefore be significantly different. This allows the scattering of light atoms, such as hydrogen, to be distinguished from scattering by heavy atoms. In addition, there can be very significant differences between the scattering strengths of different isotopes. The scattering of neutrons by nuclei depends on their nuclear spin states. There is a deviation in the scattering lengths of the nuclei from the mean value due to the differing spin states of each individual nucleus. Coherent scattering occurs if all of the nuclei in the scattering system have a scattering length equal to the mean value. In this case, scattered waves from each pair of nuclei are in phase, resulting in interference effects. This indicates that the relative positions of the atoms are the main reason for using coherent scattering. Nevertheless, besides coherent scattering, there is a contribution due to the random distribution of the deviations of the scattering lengths from their mean value, which is known as incoherent scattering. Incoherent neutron scattering methods are also used to measure hydrogen content at low concentrations. As hydrogen is dominated by the incoherent contribution, it can be distinguished from deuterium, which can be used for hydrogen-containing adsorbents. In situ hydrogenation and dehydrogenation create diffraction patterns that are determined for a series of different hydrogen concentrations, allowing for the determination of intermediate concentrations of hydrogen. In situ neutron powder diffraction studies can be carried out on both hydrides and microporous adsorbents. In the microporous materials, the refinement of structural models using the Rietveld Method allows the positions of adsorbed hydrogen molecules to be determined at low temperatures. Small-angle X-ray scattering and small-angle neutron scattering can be used to quantify microscale or nanoscale density differences in an adsorbent's molecular arrangements near the surface. The spatial inhomogeneity of the hydrogen concentration in metal hydrides, the lattice strain and dislocation formation that accompanies hydride phase formation, and the porosity of hydrogen adsorbents are detectable hydrogen storage-related features recognized by small-angle X-ray scattering and small-angle neutron scattering. Materials include a wide range of porous substances related to hydrogen storage applications, including porous silica, zeolites, and MOFs [11].

6.1.5 Chemical and Electrochemical

Methods like chronopotentiometry, a galvanostatic technique, can be used to study chemical reaction mechanisms and kinetics during ceramic adsorption. In this technique, the current at the working electrode is held at a constant level for a period of time while the working electrode potential and current are recorded as a function of time. The chronopotentiometry method utilizes a three-electrode system, where the potential is measured as a function of time by applying electrical current. The current is applied between the counter (auxiliary) and the working electrode, and the potential changes of the working electrode are measured over time. Since the current does not pass through the polarized reference electrode, the auxiliary electrode should be used to apply the desired current to the working electrode [12]. The relationship between transition time (t) at zero concentration of the oxidizing species and maximum concentration of the reduction, which indicates the endpoint of the charging or discharging process, is calculated via the Sand equation as follows:

$$t = \pi D \left(\frac{nFA_{\text{electrode}}C}{2i} \right)^2 \tag{6.14}$$

where n is the number of transition electron, F is the Faraday constant, D is the diffusion coefficient, A is the surface area of the electrode, C is the concentration of electroactive species, and i is the applied current.

6.2 FUTURE OF CERAMICS AS HYDROGEN STORAGE SYSTEMS

At the current stage, it is fair to declare that advanced ceramic materials may provide hydrogen storage solutions for hydrogen batteries. The future progress of nanoporous and mesoporous ceramic materials for hydrogen storage adsorption (storage as compressed gas or as liquid in a tank) depends on their reliability and reproducibility, as well as on their cost and affordability. While established ceramic hydrogen adsorbents, such as carbon allotropes and aluminosilicates, are widely researched and developed for physisorption and chemisorption mechanisms, the concept of ceramic encapsulation/confinement as hydrogen adsorbents is gaining ground due to newly introduced production methods and the high storage capacity and regenerability of MOs, MMOs, and oxide ceramic composites. Some of these oxides include ceria, germania, vanadia, magnesia, alumina, titania, and yttria in pure oxide, mixed oxide, and spinel forms. The available information shows that they are swiftly fulfilling DOE requirements for an acceptable hydrogen storage system in the coming years for fuel cell, domestic, and transportation applications. The cost of ceramic materials is not solely determined by the material itself but is influenced by a combination of factors, including (i) Synthesis and Sintering Processes: Complex manufacturing processes and high-temperature sintering requirements can significantly increase production costs; (ii) Purity and Composition: High-purity materials and specific dopants or additives can raise the price, but they may also enhance performance or enable unique applications; and (iii) Scale of Production: Larger production volumes can reduce per-unit costs through economies of scale, making advanced ceramics more accessible for widespread use. Therefore, future developments in manufacturing techniques and the possible application of waste materials as precursors can significantly decrease production costs.

REFERENCES

1. Nijkamp MG, Raaymakers JEMJ, van Dillen AJ, de Jong KP. Hydrogen storage using physisorption – materials demands. *Applied Physics A*, 2001, 72(5): 619–23.
2. Kim S, Kim D, Kim J, An S, Jhe W. Direct evidence for curvature-dependent surface tension in capillary condensation: kelvin equation at molecular scale. *Physical Review X*, 2018, 8(4): 041046.
3. Ladavos AK, Katsoulidis AP, Iosifidis A, Triantafyllidis KS, Pinnavaia TJ, Pomonis PJ. The BET equation, the inflection points of N_2 adsorption isotherms and the estimation of specific surface area of porous solids. *Microporous and Mesoporous Materials*, 2012, 151: 126–33.
4. Yuksel N, Kose A, Fellah MF. A Density Functional Theory study of molecular hydrogen adsorption on Mg site in OFF type zeolite cluster. *International Journal of Hydrogen Energy*, 2020, 45(60): 34983–92.
5. Dehghanian HA, Hosseinabadi N. The hydrogen storage capacity of Al–Cu alloy with permeable alumina hydrogen permeation barrier (HPB) applied by plasma electrolytic oxidation (PEO). *International Journal of Hydrogen Energy*, 2022, 47(11): 7339–50.
6. Li Q, Lu Y, Luo Q, Yang X, Yang Y, Tan J et al. Thermodynamics and kinetics of hydriding and dehydriding reactions in Mg-based hydrogen storage materials. *Journal of Magnesium and Alloys*, 2021, 9(6): 1922–41.
7. Ting VP, Ramirez-Cuesta AJ, Bimbo N, Sharpe JE, Noguera-Diaz A, Presser V et al. Direct Evidence for Solid-like Hydrogen in a Nanoporous Carbon Hydrogen Storage Material at Supercritical Temperatures. *ACS Nano*, 2015, 9(8): 8249–54.
8. Chukanov NV, Aksenov SM, Rastsvetaeva RK. Structural chemistry, IR spectroscopy, properties, and genesis of natural and synthetic microporous cancrinite- and sodalite-related materials: A review. *Microporous and Mesoporous Materials*, 2021, 323: 111098.
9. Zhang X, Sun Y, Xia G, Yu X. Light-weight solid-state hydrogen storage materials characterized by neutron scattering. *Journal of Alloys and Compounds*, 2022, 899: 163254.
10. Motazedian M, Hosseinabadi N, Khosravifard A. The ceria – Germania solid oxide hydrogen storage hollow porous nanoparticles. *Materials Chemistry and Physics*, 2023, 307: 128100.
11. Sheppard DA, Maitland CF, Buckley CE. Preliminary results of hydrogen adsorption and SAXS modelling of mesoporous silica: MCM-41. *Journal of Alloys and Compounds*, 2005, 404–406: 405–8.
12. Gholami T, Pirsaheb M. Review on effective parameters in electrochemical hydrogen storage. *International Journal of Hydrogen Energy*, 2021, 46(1): 783–95.

Index

For Product Safety Concerns and Information please contact our EU representative GPSR@taylorandfrancis.com
Taylor & Francis Verlag GmbH, Kaufingerstraße 24, 80331 München, Germany